METHOD AND PRACTICE IN ARISTOTLE'S BIOLOGY

Michael Boylan

UNIVERSITY PRESS OF AMERICA

LANHAM • NEW YORK • LONDON

University Press of America,™ Inc.

4720 Boston Way
Lanham, MD 20706

3 Henrietta Street
London WC2E 8LU England

Printed in the United States of America

Library of Congress Cataloging in Publication Data

Boylan, Michael.
Method and practice in Aristotle's biology.

Bibliography: p.
Includes indexes.
1. Biology—Methodology—History. 2. Biology—Philosophy—History. 3. Aristotle—Science. I. Title.
QH324.B68 1983 574'.092'4 82-23708
ISBN 0-8191-2952-6
ISBN 0-8191-2953-4 (pbk.)

All University Press of America books are produced on acid-free paper which exceeds the minimum standards set by the National Historical Publications and Records Commission.

TABLE OF CONTENTS

Preface

This book is intended for readers who want to understand and evaluate Aristotle's philosophy of biology and his actual biological investigations. The scope of treatment cuts across several academic disciplines so that I have tried not to assume that the reader have any specialized knowledge of Aristotle, biology, the history of biology, or philosophy of biology either in the questions discussed or in the literature about them. The Greek has generally been transliterated and in most every case the arguments being examined should be intelligible to the reader without Greek.

What makes Aristotle's biological works so stimulating is that they incorporate a large number of different concerns into a single treatment. Aristotle, I believe, correctly delineates some of the most crucial aspects of biological methodology which have significance not only to those interested in history and classical philosophy, but also to those interested in the problems of contemporary biology. Indeed, it is the very structure of biological investigation to which Aristotle speaks and the mode of this structure's construction that condition and determines the outcomes of those investigations. Thus, the modern biologist and philosopher of biology will find Aristotle speaking to them too, once his methodology is properly understood.

In this book I have tried to illuminate the significance of Aristotle as a philosopher of biology and as a biologist. This is my general purpose, and it requires first some background into the scientific climate in which Aristotle found himself. It is impossible to appreciate fully the large strides Aristotle made without also recognizing the types of methodology and practice which went before him. For in Aristotle's own lifetime, great advances were being made in medicine and natural philosophy, and a degree of sophistication was achieved that is truly remarkable, considering that only two hundred years before, Greece was still in a pre-scientific age. The development of methodological principles is difficult to trace. One approach which is useful here is to trace when reason (<u>noēsis</u>) parts courses with sensory experience (<u>aisthēsis</u>). When these are separated from understanding (<u>nous</u>), their relative positions in creating explanatory theories are fraught with difficulties. Aristotle offers his own solution to this dilemma through a general theory of induction discussed in the _Posterior Analytics_.

In my first chapter, then, I outline the relevant context in as much detail as the limited space allows. This treatment is in no way intended to be exhaustive, but rather is suggestive of a particular way to understand both the pre-Aristotelian natural philosophers and the pre-Aristotelian biologists.

This context sets the stage for Aristotle to provide unique solutions to methodological dilemmas and for this new methodology to influence the way he actually sets about practicing biology. Now there are many commentators on Aristotle who feel that these two steps are not connected. Such persons argue that Aristotle's biological investigations contradict his methodological dictums. Implicit in such a charge is that Aristotle's method for biological science is inadequate for actual biological investigation. It is my contention that such objections are unwarranted. To support such a claim it is incumbent upon me to outline in detail: first, Aristotle's methodology of biological science, and second, the way in which his actual investigations conform to the specifications of his methodology.

Thus, in Chapters Two and Three, I examine in some detail the principal treatise on biological methodology, _De partibus animalium_ 1. Within this short treatise numerous important questions are addressed and I endeavor to discuss the significance of these. From these chapters the two principal modes of biological explanation are outlined. These are: A. Teleological explanation (TE), incorporating the formal and final causes, and 2. Material or Mechanical explanation (ME), covering material and efficient causes. These are developed in Chapter Three under the broad concept of Necessity, which is the vehicle by which Aristotle presents these.

In Chapter Four, a representative sampling of Aristotle's actual biological investigations is brought forth and examined in light of Aristotle's method. The detail and accuracy of most of the observation and explanation found here are remarkable.

Finally, in the last chapter, some consequences of and possible objections to Aristotle's biological science, as well as my reconstruction of it, are examined as the applicability of this methodological system to modern biology is made clear.

Also, contrary to what many think, it is contended that the methodology in *De partibus animalium* 1 is generally consistent with Aristotle's overall philosophy of science, as I outline in Chapter One.

It is hoped that some of these considerations will speak to modern philosophers of biology and those interested in contemporary problems in methodology, suggesting interesting directions for them as well as illuminating the work of one of history's greatest philosophers.

This project began as a dissertation and has grown from the original conception. Special thanks go to Alan Gewirth, Arthur Adkins, Ian Mueller, Anthony Preus and David Hull for their help at various stages of the process. Each went through at least one entire version of the manuscript and offered many very helpful suggestions. I would also like to thank an anonymous referee for helping me improve several problem passages, and would like to add my appreciation for the help rendered in the preparation of this manuscript for publication by Ann Mallinger, Bob Smith, David Guercio and Richard Halverson.

Lastly, I owe a great thanks to my wife, Rebecca, who has offered her advise on a number of substantive and formal issues which have improved the manuscript.

CHAPTER 1

THE CONTEXT

1.1 PRE-ARISTOTELIAN NATURAL PHILOSOPHY

What are the parts of a scientific theory? Clearly there are at least two: the *explanans* and the *explanandum*. A person is presented with some sensory data and the phenomena are explained in a theory. Thus one has the objects, the sense experience and the explanation of that experience. By explanation, then, one presumably wants to discern the true nature of what is presented to him. This "discerning" may take the form of some categorization of the phenomena and/or an aetiology to offer the *causes* of them. Presumably then, one needs rules, a *methodology*, to guide in the construction of these aetiological theories.

All this may seem straightforward and general enough from our perspective today, and yet in Homeric Greece, science and scientific explanation had not developed to this point. This can be seen from a brief examination of the words for reason and reasoning, *nous* and *noein*.{1}

In Homer, *noos* seems connected with the sensory faculty of sight. Just like words such as *idein* and *gignōskein*, which also translate as "to know" or "to understand," in fourth century Greece, *noos* in Homer connoted an evaluative attitude in certain contexts, while in others it merely signified a mode of apprehending what is presented to the senses. This note of evaluation is often portrayed in literature as gods disguise themselves to mortals. Somehow it is *noos* which enables the mortal in question to penetrate beyond the surface appearance to apprehend that he is speaking to a god.{2}

THE CONTEXT

One obvious consequence of this is that there was a union between sense experience (particulaly sight) and some kind of "rational" awareness. The word "rational" must be viewed in inverted commas since reason had not yet become critical.{3} That is, reason had not yet acquired the role of making judgments about the sense experience presented to it, nor of evaluating its own operations. Indeed, *noos* had not even taken on the specialized role of "reason" as yet.

Now the precise time when this bifurcation occurred is rather difficult to determine because there are few sustained scientific discussions (or places where *nous* is used) available. The texts that do remain are usually only fragments. Further, a complete understanding of the pre-Socratics requires a detailed discussion of their intellectual and social environment in order to appreciate the problems which faced them and how they viewed their roles as natural philosophers. However, it is clear that after Parmenides of Elea (circa 480 BC){4} there was this distinction between sensory impressions and reason which makes judgments about them. For it is with Parminedes that the first major division between ancient thinkers can be drawn, viz., between those who held *nous* as the highest principle and those who favored *aisthēsis*. These two dispositions will be labelled, initially, uncritical rationalism and empiricism.

Now, invariably, discussions on empiricism often become confused because of the terminology invoked. Writers who mistrust theories and the generality they provide because of some skeptical predisposition will be called uncritical empiricists. The Hellenic school of the Empirics would fall under this heading.{5} The second group (which asserts the necessity of both observation and rational input in any explanation to varying degrees) will be called critical empiricists. Aristotle falls under this second group. Finally, the last group will be called uncritical rational or *a priori* theorists. This last group often found itself in opposition to the uncritical empiricists who mistrusted their use of logical hypotheses that often strayed far away from sensory data.

These kinds of distinctions are given various labels by different writers, but it is hoped that this terminology will make clearer some of the distinctions between these modes of physical explanation. They are also terminologically closer to the way in which Aristotle refers to the practices of these physicists (*phusiologoi*).

THE CONTEXT

Now there is a sense in which the so-called 'empiricist' and 'rationalist' (critical and uncritical) are not as far apart as is often portrayed. This can be illustrated by two figures from the Seventeenth Century. Francis Bacon, though enthusiastically endorsing experiment and enumerative induction as a basis for knowledge, was careful to modify these results with a rational method. Descartes, though he spends a good deal of time on a rational method and a rational metaphysics (Rules for the Direction of the Mind, Meditations on First Philosophy and Discourse on Method of Rightly Conducted Reason), used this base for his own science in which he conducted experiments. It is true that there could be no rationalism without some sensory input and no empiricism without a rational structure in which the information is organized and interpreted (Aristotle, himself at one point describes induction as partly a rational process, APr. 68b 15). However, it is incorrect to view the empiricist (even the uncritical empiricist) as recording only what he sees, and the uncritical rationalist as wildly speculating about dreams and phantasms. Both approaches emphasize different aspects to be found in all scientific explanations. No exposition about scientific method would be accurate which attempted to draw too radical a distinction between the two. And yet there are two ways in which one can talk about the distinction between uncritical rationalism and the two forms of empiricism even though such a distinction may only admit subtle differences in degrees of usage.

The first way involves scientific discovery. In discovery one has phenomena and an explanation. Now to explain some phenomena, a theoretical move is necessary since creating an account involves contemplation about experience. This is common to all explanations. Each has sensory input which is rationally deliberated, however; the key question at this point is, "Where does one get such an explanation?" Explanations, when deductively systematized begin with certain postulates or first principles. These postulates lay down some basic conceptions about the explanandum. The empirical thinker seeks (as much as possible) to derive his first principles from the world about him chiefly through his own experience (though common opinion cannot be discounted either). The uncritical empiricist pretends to stop there. The uncritical rationalist seeks to explain phenomena by imposing a framework on reality which has its genesis in his own reflections on the problem.

Some postulates are, in principle, impossible to obtain through first hand observation, viz., those which concern microscopic phenomena or some astronomical phenomena. In these cases one must bring in a second criterion: verification. Observational verification can be either direct or indirect. In the direct observational mode the observer can verify or falsify some explanation (or corollary) through an experiment or continued observation. Indirect observational verification involves viewing results that must occur if the theory is correct (thus testing the explanatory power and predictive power, as is the case in sub-atomic physics).

Though explanation necessarily involves theoretical and observational elements, this is not to say that these terms, 'uncritical rationalism,' 'critical empiricism' and 'uncritical empiricism' are meaningless. Despite the fact that there are many problems in trying to make distinctions between these terms, there is certainly (at least) a difference in emphasis between the two extremes: uncritical rationalism and uncritical empiricism. It is for this reason that the following sections use these strains: roughly to judge the natural philosophers which formed the scientific context for Aristotle. The median point, critical empiricism, can itself be weighed either to the side of uncritical rationalism or to the side of uncritical empiricism so that there are no hard boundaries between these modes of theorizing. But various movements and progressions toward the center from the extremes is instructive in understanding what Aristotle was doing.

Now it is not all together clear to many commentators on Aristotle whether he was a critical empiricist or not.{6} Ingemar Duering believes that Aristotle was not an empiricist: "Yet he (Aristotle) was perhaps to a far greater extent than we sometimes are inclined to believe, a desk work scholar . . . he regarded phusikē as a theoretical and not an empirical science."{7} This view seems to be supported by the French writer G. Pouchet, "Aristotle was never an experimenter."{8} These writers believe that Aristotle only acquired his knowledge of nature second hand from butchers and fishermen. They make Aristotle out to be a theorist who only collected information from others. Now this, in one way is a strong charge. In another way it is not. Certainly, as a scientist, one must rely upon a body of knowledge to be given to him. One cannot, and should not, make every observation himself. If each scientist had to start from scratch and form his own theories from the beginning, then there would be no progression of science or technology. It is necessary to accept a certain body of information. On the other hand, one must do a certain amount of first-hand work himself so that he can understand the full significance of the phenomena before him. This is especially so when one is literally forming a new scientific discipline as Aristotle did with biology. Because of this it will be important to ascertain just how much work Aristotle actually did do in the field. For even though one can be a critical empiricist and be largely dependent upon reports of others, in the field of biology (especially one who is its founder), the true critical empiricist would want to be constantly checking his theories against the data and vice versa.

It is unclear just when these types of distinctions between the critical empiricist and its uncritical counterparts first appeared, but an answer to this question can perhaps be found by returning to the analysis of nous begun earlier. Now some believe that it was Thales (circa 585 BC) who first caused the bifurcation of nous into rationality and sense experience. Others give the nod to Xenophanes (circa 520 BC).{9} However true this may be, the texts are too scanty to make a judgment. However, in the case of Heraclitus of Ephesus (circa 500 BC) these two tendencies are at least heightened, if not actually separated. He says, "Bad witnesses are eyes and ears to men if they have not the understanding of their language in their souls"(Fr. 107).{10} Vision and other forms of sense perception may be useful for knowing (Fr. 55, 101); though it may be deceptive (Fr. 56); and is not sufficient for understanding (Fr. 40). This is possible because one may see or hear and not understand (Fr. 17, 19, 34).

What is one to make of this? There has been much discussion about what exactly Heraclitus is doing.{11} Is he putting either aisthēsis or nous on top? There are three fragments in which nous is mentioned: B40, B104 and B114. In the first two of these the idea is asserted that nous is only possessed by a few.{12} The third fragment refers to a common basis for all human knowledge.{13} This seems to accord with his discussion on logos. This logos{14} is the basis for all knowledge and is identified spiritually with the Divine and temporally with fire.{15}

It seems to me that in these passages one has aspects of a separate nous and aisthesis but that the separation is not yet fully complete. There appears to be too much interplay between the two. But if one were to point to tendencies, Heraclitus is leaning in the direction of the supremacy of logos and nous. This is because, on balance, Heraclitus does not ultimately want to trust his senses (the two modes are still united though the division has begun through all the cryptical sayings).{16} Heraclitus does not ultimately want to trust his senses. He is forced to acknowledge their input because the separation of nous and aisthēsis has not yet become complete. There are in Heraclitus the seeds of a growing skeptical attitude concerning the place of sensory knowledge.

With Parmenides this split is finally completed. This split is represented respectively in the two halves of his famous poem. The first half discusses the truths of nous, noein and noēma, {17} while the latter half represents the "seeming" (doxa) of the empirical world. This contrast follows the nous/aisthēsis dichotomy as outlined earlier. Even ancient commentators such as Simplicius declared that Parmenides was moving apo ton noēton epi ta aisthēta (Phys 30,14). This does not mean that nous is absent from the second part of Parmenides' poem (as B16, 1.2 and 4 attest), but there is clearly a different emphasis in the second part of the poem.

There has been a great deal written on the contrasts and similarities between these parts, but the most reasonable reading views the first half as affirming the truth and dominion of reason.{18} This contrasts with the second half of the poem which represents the world as most people see it. This world of appearance is filled with opposites because the data acquired by these means is essentially false (or at least subservient to that of reason). The mode of exposition in the poem is extremely subtle and these themes concerning the relative places of nous and aisthēsis are underscored through ambiguity and irony.{19}

Thus beginning with Parmenides there follows a series of philosophers who tend toward uncritical rationalism. Nous is elevated and seen to be sufficient in itself to apprehend the truths of what "is." Following in this tradition are Zeno of Elea and Melissus of Samos (circa 445 BC), and Philolaus of Croton and Eurytus of Croton (circa 410 BC). Finally in the fourth century one of the most famous advocates of this position was Plato of Athens (circa 360). Plato ultimately mistrusted the senses. This information was relegated to the lower half of the divided line. Any physical account by its very nature could be no more than a likely story (eikota logon, Timaeus, 48d). Because of this, one's attitude toward a physical theory would be that a theory may demonstrate certain rational relationships which, if they are true, make the theory a good one. The physical details are relatively unimportant since they only represent reflections of something more basic (Republic 510a).

Perhaps one can get an idea of Greek scientific method from examining the third level of Plato's divided line, *dianoia*. As a rational idealist, Plato put mathematics on this portion of the line. Like Parmenides (and the Pythagoreans), Plato especially honored the place of mathematical and so placed the mathematical method at this level. The method is that of laying down principles and following deductively to conclusions. One asserts a postulate (*hupothesis*), such as odd and even, as if it were true and proceeds to the end of his inquiry (*Republic* 510c). These hypotheses are set down without any further account (such as the highest level would have). The mathematician also uses images to represent his "squares and diagonals" (*Republic* 510d).

According to Plato these conclusions are tentative because the *archai* have not been given a *logos*. But the mathematical method is still highly respected by Plato, occupying the penultimate position in his epistemological hierarchy. The criterion for acceptability here is that the system be internally consistent (i.e., the premises do not contradict each other and, if accepted, cause one to necessarily accept the conclusion). This is as far as the mathematician qua mathematician need concern himself. The truth of the conclusion depends upon a justification of the *archai* (generating principles) which is in the province of the philosopher.

This emphasis upon internal consistency above all other considerations has been discussed by A. Wasserstein who believes that all early theories were of this nature.{20} He interprets the famous dictum "*sōzein ta phainomena*" as an activity of finding appropriate postulates and deducing their results. "Appropriate" here refers not to 'true' or 'real,' with respect to the world we see, but in accord with certain attractive notions such as perfect circular motion, symmetry and the like. The *phainomena* thus become problems that the scientist wishes to solve. "*Sōzein ta phainomena*" becomes equivalent to "solving the problem with convenient or attractive postulates." An example of this can be found in Ptolemy's reasons for adopting the geocentric hypothesis over the heliocentric one.

> It is more reasonable to adopt the hypothesis of eccentric movement because it is simpler and because it achieves its object by means of one supposed movement, not two (two at least are of course needed for an epicyclic system).{21}

The competing theories here are eccentric movement and epicyclic movement. The eccentric movement can account for irregularities in planetary movement by moving the earth away from the center of the planetary orbits and constructing a deferent which rotates in the same direction as the planets

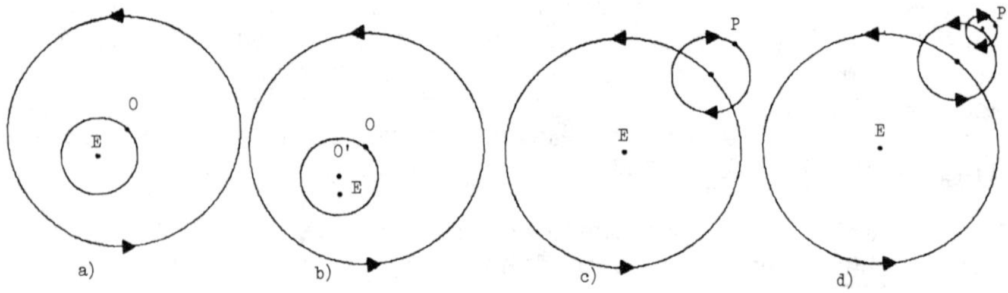

FIGURE ONE: The Motion of the Planets

(figure 1, a & b). This explanation posits only one movement, i.e., counterclockwise movement. In the epicycle system two motions are required (figure 1, c & d); a clockwise motion and a counterclockwise motion. Since and c, and b and d are equivalent, Ptolemy chooses the eccentric account; it is simpler. At first glance this would merely be an illustration of invoking that time honored doctrine of simplicity which has been much discussed and debated in recent years (see Chapter Two, n. 3). But upon closer examination this can be seen to be different in one important respect: the realistic assumption. When simplicity is invoked today it is to adjudicate between essentially equivalent real explanations. If one theory x was simpler than y, but was also known not to be in accord with the "real" structure of things, then x would be rejected no matter how simple it was. Simplicity *alone* is not sufficient for choosing between theories—even mathematically equivalent theories. Mathematical equivalency says nothing about what *is*, and it is this realistic variable which is paramount to scientific discovery. Ptolemy never considered this. It seems to have been unimportant to him. The key here was economy and had nothing to do with what was *really* the motion of the planets. Physical relaity played no part in his decision. "For Ptolemy the choice is simple: of two hypotheses, different in their structure but equivalent in their mathematical results, he would choose the more economical one" (Wasserstein, p. 55).

This procedure is taken up in Proculus as well. He wants to show that the movements of the planets are pure abstractions; they exist only in the mind of the astronomer.{22} Indeed, Proclus thought the motions to be irregular in nature, but despite this he gives assent to the assumptions of the uniformity of motion (believing them not to be in accord with nature) for the purposes of the problem at hand which are essentially mathematical and geometrical.{23}

The phainomenon here becomes a logical problem with difficulties (aporiai) which must be solved. The simplest, most economical road to the solution is the one which is adopted. Sometimes this is carried to the point of discarding data known to be correct for alternatives which fit more easily into the calculations. Aristarchus of Samos, one of the most competent ancient astronomers, does just that when he chooses two degrees to be the angular diameter of the moon for his calculation (it should be one-half degree) just because they fit better.{24} Apparently Aristarchus was more interested in demonstrating his method than he was in arriving at a correct calculation. This strain of treating phenomena as problems or common opinion lends itself to uncritical rationalism.{25}

This emphasis upon internal consistency above other considerations may have grown from the mathematical bent of all these philosophers. They concentrated on solving particular problems and constructed systems with certain requirements in mind which would yield the solution. This solution only had to satisfy the requirements of the system they had. This might be fine for mathematics, but for natural philosophy it tended toward an instinct to the *a priori* or uncritical rationalism.

In Empedocles we find a writer who seriously tries to form a methodological basis for uniting *noēsis* and *aisthēsis*. This is moving in the direction of critical empiricism. Empedocles makes it clear from fragment 3 that sensory knowledge is to be valued:

> But come, consider with every effort how each thing manifests itself, neither holding sight in higher stead than hearing, nor loud-sounding hearing above evidence of the tongue, nor holdback trust from any other limb, wherever there is a path for understanding, but think on each thing in the way it manifests itself.

However, it is dubious that such a relationship actually existed. What is important is that while holding sensory knowledge in high regard he posits a "materialist" theory of the universe and its single and progressive evolution through the influences of Love and Strife;{26} he also embraces a strong religious, a priori strain which seems to contrast with the former. This seeming disparity has bothered readers of Empedocles. For in his poem, On Nature, his doctrine of the four elements, earth, air, fire and water, is combined with the non-spiritual motive forces of Love and Strife. These are sufficient to give an account of the universe in strictly materialistic terms. However, in his poem Purification, he seems an other wordly seer who exhorts all to purify themselves according (perhaps) to "Orphic" or Pythagorean mystery cults. The Empedocles of this poem appears to be an uncritical rationalist.

A partial answer to this dilemma might be that Empedocles is in the context of an emerging scientific tradition in which uncritical rationalism has many adherents (as does uncritical empiricism), and that he is taking the first steps toward unifying the two under the guise of critical empiricism.

At around the same time, Anaxagoras was also attempting to perform the same task. Like Empedocles he tried to offer a theory which would provide some integration of these emerging modes of explanation. Against the uncritical empiricists he says:

> From the weakness of our senses we cannot judge the truth. (21) Appearances offer a sight of the obscure. (21a) {28}

It is not possible from observation alone to get at truth. This is what the first quotation seems to be saying: it is not possible (ou dunatoi) to judge the truth from our senses. If this is so, from whence does one judge the truth? Perhaps the key word is krinein. This could mean that the senses alone cannot decide or judge the truth correctly. They only offer a sight or something obscure or unclear (21a). This unclear image must be put together in the mind of the perceiver.{29} The phenomena do not put themselves together. They must be arranged and judged by a mind. Read in this way the passage is not as anti-observational as it first might appear to be.

From this it seems clear that observation alone will not yield truth. Anaxagoras believed that one must go beyond the senses as he posited a theory which could account for material composition. This is the famous theory of "seeds." These seeds were both supposed to be homogeneous (homoiomerous){30} and yet they contained principles for all other matter within them:

> ... and the seeds are of an indefinite number and are unlike each other. For none of the others are alike to each other. Because of this we must suppose that all things are in this whole (Fr. 4).{31}

This apparent problem can be resolved in any number of ways,{32} but what seems clear is that Anaxagoras (like Empedocles before him){33} has tried to offer a physical account which goes beyond mere sensory knowledge and tries to construct a single theory which will have wide explanatory power.

Anaxagoras did improve on the Empedoclean doctrine of knowing "likes" by "likes" (Fr. 109) by postulating perception through contrast (Theophrastus, de sensu 27 ff., DK 59a 92). Surely perception contains both elements. However, at this point it is not too profitable to ask how the categories of critical empiricism over and against uncritical rationalism and empiricism fit in here since our data are too sketchy. One thing is clear, though, that theories of perception were still underdeveloped and what seemed important to these thinkers was offering an explanation in the modes which they were developing.

The same sorts of comments might be made about Leucippus and Democritus. The effort of theorizing through a small number of postulates is, in one way, heading in the right direction, however; just how these theories tie in with the problem of noēsis and aisthēsis is unclear (from the texts available). It seems that these two philosophers were responding to the Eleatic dilemmas which, themselves, had definitely made the split between noēsis and aisthēsis. Leucipus and Democritus tried to respond in such a way that the appearances might be saved and the "counter-aesthetic" monism of the Eleatics could be rejected. While Empedocles and Anaxagoras still accepted certain Eleatic assumptions, e.g., on the impossibility of the void, Leucippus and Democritus come out and offer a clear alternative. This alternative appears to have been worked out in some detail (another advancement upon earlier thinkers). For example, if we can trust Theophrastus' account in his de sensu 49-83, Democritus offered a comprehensive explanation of the mechanism of the different senses in terms of atomic theory.

This points to theory integration and systemic application which are indicators of a trend toward sophistication and scientific development.

For further refinement on the rules, by which theories should be constructed in order to avoid error and arrive at the truth, and to continue this trend towards system building and theory integration, we have to move first to the medical writers and then to Aristotle himself.

1.2 PRE-ARISTOTELIAN BIOLOGY

In biology one is tied to observational facts in a way that is different from the other natural sciences. From its beginnings in medicine, the physician could not avoid his data: the patient and his symptoms. There were two traditions in ancient Greece: the magical and the empirical. The magical tradition is mentioned by Plato in the Charmindes 155e, Symposium 202e-203a, Theaetetus 149c-e, and the Laws 666b-c, 798. In these passages charms and incantations used for healing are alluded to. In Peloponnesian War 2.47, the skills of the physician as well as those of magic are needed to effect a cure. Strabo was correct in Geography 14, 657, when he says that Greek Medicine was born in the temples.{34}

There were two major schools of medicine at this time on the neighboring Cos and Cnidos.{35} It is in one respect hard to draw too sharp of a distinction between these since both are empirical and, if we are to believe Galen, both represented major forces in early medicine.{36} But the Cnidian writers tended to be less concerned with offering causal theories for diseases, but tried to stick as closely to the symptoms as possible.

> Another disease: an intense head pain; vomitting of bile, difficulty urinating and delirium. The patient has pain from an excess of heat in the brain. He is delirious when the blood in the brain is heated and is caused to move violently about by bile or by phlegm. He vomits bile (this liquid being moved in the body and the head), because of the heat, draws itself inward; and on the one hand vomits the thickest and on the other the thinnest is drawn in. One urinates by the same causes just as has been said earlier . . . by the seventh day he often dies or if he makes it to the seventh he can expire on the ninth or the

eleventh at least. (Epidemics II, Littre 7, p. 11, 26).

In this passage the author describes symptoms and gives a brief cause for them. The recommended treatment of hot compresses, barley soup and white wine are in no way causally connected with the disease. They are merely cures which have been tried before and found to have worked presumably through trial and error (since there is no characteristic theory why one treatment is preferred over another). The treatments prescribed are such as to be applicable in every case. That is to say that the symptoms dictate the cure to be used.{37} It is almost as if the symptoms were themselves the disease.{38} The external signs are themselves what must be eliminated. When seen in its extreme form in Epidemics III, aetiology is eliminated altogether.

> The mortification of the gangrene. If the gangrene mortifies itself there is a head pain and frequently a scratchy throat; the sick limb loses sensation, a feeling of cold comes to the head and the affected limb sweats. He suddenly loses his speech and blows blood from his nose as he becomes pale. If the disease takes hold of the patient with a weak force, he recovers the discharged blood. If the disease takes him with a strong force, he dies promptly. In this case one induces sneezing by pleasant substances; one evacuates by the upper and lower. Alternatively those odors will be a little active. The soup will be light and hot. Wine is absolutely forbidden. (Epidemics III, Littre 7, p. 123).

All one is left with are symptoms and treatment. But when one practices medicine in this way there are severe restrictions. For the disease is seen as a collection of symptoms. The cure can only be guessed at unless it has been previously written down in a manual. When a physician is confronted with a novel disease he must find a similar set of symptoms and use that treatment. This aspect of the Cnidian method brought harsh rebuke from Galen.

> The point is that they [the Cnidians] looked at the varieties of symptoms which change for many reasons and failed to consider the specificity of the dispositions, as did Hippocrates, who used for their discovery a method, only by using which, one can find the number of diseases . . . . Hippocrates censures the Cnidian physicians for their ignorance of the genera and species of diseases, and he points out the divisions by which what seems to be one

becomes many by being divided. (*Corpus Medicorum Graecorum* 5.9.1 p. 121-2; *Claudii Galeni De Placitis Hippocratis et Platonis*, ed. I. Mueller (Lipsiae, 1874), p. 776).

What was it that made the Cnidians different from the Coan writers? This can be found by examining the two steps in any medical practice: prognosis and treatment. In the Coan work *On Prognosis* the writer suggests that prognosis consists in knowing the patient's condition in the past, present and future. Now how could a physician know this? Well, this could also be put into a handbook as a case by case description and memorized. But the problem lies in that each case *is* individual and the physician relying on such methods would surely be forced to give more general answers while being a slave to what had been described about the case (as the Cnidians were).

The alternative is to have a theory about what caused the diseases in the first place. This allows a distinction between disease and symptom so that a large number of similar cases might be grouped together into classes of diseases. The physician need only learn about these general cases and fit each patient into the appropriate, general category. Thus a physician might take the individual patient's constitution into consideration when deciding how certain symptoms of a disease group might be expressed. This emphasis on the individual is distinctly Coan.{39} Now a physician, when confronted with a new set of symptoms that are beyond his experience and learning, needs only to try to determine whether the disease falls within a larger category. In this case the Coan physician may have a possible cure derived from his general theory, while the Cnidian physician must begin a new trial and error process.

Obviously treatment under the Coans no longer is "hit or miss" but results from the supposed causes of the underlying disease. The novel case, once diagnosed according to its causes, can be treated relative to those expressed causes.

The real difference here is that the Cnidians are uncritical empiricists. They have no appreciable theoretical structure by which their practice might progress. They cannot move beyond the phenomena (as problems or common opinions, *endoxa*){40} because they are slavishly tied to it. The Coans, on the other hand, have tried to incorporate something in their empiricism. This is an overarching theoretical structure. But how is one to determine this structure? Does one randomly set down hypotheses and derive deductive conclusions for a "likely story" as Plato did--or as the writer of *Ancient Medicine* accuses other natural philosophers of doing?{41}

> All who, attempting to speak or write on medicine, have assumed for themselves a postulate as a basis for their discussion (_hupothesin autoi autois hupothmenoi tō logō_)--heat, cold, moisture, dryness, or anything else that may fancy--who narrow down the causal principles of diseases . . . obviously blunder.{42}

Now _hupothesin_ here can be taken to be the same laying down of postulates discussed in the previous section. This writer's statements here indicate that he is against the method employed by the astronomers and other non-biological natural philosophers who accept arbitrary principles and then deduce results. This would appear to be a conceptual shift in method, however; the categories offered by the author to account for disease, the bitter, the sweet, the acrid, the astringent etc., are equally as arbitrary as those which he attacked in his opening sentence. Again in Chapter One of the _Nature of Man_, Polybus assaults the practice of assigning such vague and random causes for things as earth, air, fire and water, while he himself later offers his own four part theory of humours: blood, yellow bile, black bile and phlegm.{43} Each are associated with two contraries: hot, cold, wet and dry; as well as a specific season: spring, summer, fall and winter. Again as in the case of the writer of _Ancient Medicine_, the categories offered fall prey to the same objections first raised by the author about his contemporaries. Though the process of laying down arbitrary hypotheses may have been seen as problematic, somehow, these writers were unable to remedy this methodological problem in their own writings which remained just as _a priori_ as those they attacked.

But what is important in all of this is an effort to form a critical empiricism which steers a middle course between the naive and unorganized acceptance of sensory data by the uncritical empiricists and the _a priori_ bent of the uncritical rationalists. This can be further illustrated by _Ancient Medicine_ 20:

> I hold that clear knowledge about natural science can be acquired from medicine and no other science . . . I mean to possess this information, what man is, by what causes he is made and similar points accurately . . . for there are many other bad foods and bad drinks which affect men in different ways . . . To take my former example cheese does not harm all men alike, some can eat their fill of it without the slightest hurt . . . but if cheese were bad for

> the human constitution without exception, it would have hurt all.

In this passage several important items are demonstrated. First, medical information must arise from data acquired from medical sources, i.e., from humans. Astronomy and other *a priori*, non relevant sources cannot provide information proper to biology. This thesis asserts a separation between the various sciences. Secondly, 'bad' and 'good' are not always absolute terms since their effects can vary according to the individual. The point here is that the "badness" does not reside within the food, but results in the interaction between that food and some particular constitution. There is also an emphasis given to interpreting the data correctly.

> . . . their art resembles divination; for diviners too think that the same bird which they hold to be a happy omen on the left is an unlucky one when on the right while other diviners maintain the opposite. The inspection of entrails shows anomlies in its various departments. (*Regimen in Acute Diseases*).

The practical reason for this is clear from the passage. If physicians want to be more respected than diviners, they must scrutinize their data. The underlying supposition is that there is a "proper" way to interpret these facts. If one looks properly and has a correct understanding of the body and of diseases, then a single course of action will become clear. This contrasts to the fragmented "case by case" approach of the Cnidians.

Following a methodology can lead one (if he is not careful) into difficulties in such interpretation of results. This problem comes up in *Airs, Waters and Places* 8, where it is suggested that there are light and heavy elements in water. The former are noble and beneficial while the latter are base and harmful. To show that there are two parts to water the writer suggests an experiment. One is to put a measured quantity of water into a vessel and let it freeze. Then one is to let the water thaw again the following day. If the water is measured again it will be found that there is much less than there was originally.{44} This is supposed to show that "freezing dries up and causes to disappear the lightest and finest part, not the heaviest and coarsest, to do which it has no power" (*Airs, Waters and Places* 8). Again the fact that there is less water after freezing and thawing than there was before does not mean that there are two parts to water, nor that one is lighter than another, and certainly not that one is base (*ponēria*). In fact, if the experimenter had been a careful observer, he would have found that freezing water makes it expand in volume--just the opposite effect from the theory's prediction. It is only in the thawing that water is lost--not in the freezing--so that if a theory is meant to be explained by this example it must be a theory about thawing.

One aspect that this example illustrates is the lack of careful and detailed observation among the Coan writers which is in contrast to the very explicit descriptions of the Cnidian school. This is not always the case, of course, but it sometimes leads to *a priori* strains such as are evidenced in *Breaths* 1 in which a very broad conceptual scheme is offered which seems to be primarily verbal in nature. The cure for hunger is food and the cure for thirst is drink. All diseases are said to be cured by their opposite. *Fever* in 7 is said to be caused:

> When the body is filled full of food it becomes full of wind too, if the food remains a long time; and it remains a long time due to its bulk it cannot pass on. The lower belly is thus obstructed. Because of this, the breath (*pneuma*) spreads throughout the body and striking the parts of the body that are mostly full of blood it chills them. These parts, where are the roots and springs of the blood, being chilled cause a shiver to pass over all the body.

This catch-all concept of *pneuma* operates in even a freer, less precise way than it did in *Airs, Waters and Places* and *Sacred Diseases*. It is invoked to explain everything in lieu of any observations or even a more precise, purely rational explanation. Even pain due to cuts owes its origin to *pneuma*, "Lacerations in all cases occur for the following reason. Whenever flesh is violently severed from flesh and the wind slips into the gap the pain is thereby produced" (*Breaths* 11).

Breaths and *Nutriment* are perhaps the least empirical of all the Hippocratic writings. By emphasizing purely verbal analysis (e.g., *Nutriment* 8, "Nutriment is that which is nourishing; nutriment is that which is fit to nourish"), the authors move toward uncritical rationalism.

Though there are many good observations in the Coan School, they characteristically are not as detailed and exact as in the Cnidian School. Even in books designed to be precise, such as

On Fractures and On Joints, the descriptions of bones and setting procedures are very sketchy. This opinion is emphasized by Dr. E. T. Withington as he mentions that two very different injuries, an ankle sprain and an ankle dislocation, are not differentiated but are presented as being the same.{45}

With such a large body of writings, sweeping generalizations are difficult to make. However, this much is clear: 1. Cnidians employed careful observations though they were tied to uncritical empiricism for the most part, and 2. The Coans, while making progress toward critical empiricism, through methodological checks upon their theoretical activity (and thus making advances over Empedocles and Anaxagoras in this respect), nevertheless, were unable to carry these impulses far enough either in their methodology or in their biological/medical practice.

1.3 ARISTOTLE'S THEORY OF CRITICAL EMPIRICISM

Within the strains of Greek science which have been traced thus far it has been suggested that there have been attempts to create a methodology which would enable theories to be strictly regulated in their execution (to guard against uncritical rationalism). This represents a trend is toward critical empiricism. Aristotle in APo. makes several key distinctions which may be of some importance here. In the syllogism, which is discussed in APr., there are premises and a conclusion. The entire syllogism consists of three terms: a middle and two extremes (APr. 1.5). In the premises the middle connects with one of the two extremes in each premise. These extremes then form the conclusion with one standing in the subject position (the minor term) and one residing in the predicte (the major term). To satisfy the requirements of scientific knowledge the method of demonstration, apodeixis, is adopted, 71b 17. Demonstration results in a certain type of syllogism. Formally the syllogism must be in the first figure (APo. 1.14).{46} Secondly, the premises of this type of syllogism must be true, primary, more intelligible, prior and better known than the conclusion and the cause of the conclusion. By "cause" here one must assume that it is knowledge of the premises' essences that "causes" the conclusion. Since the premises' essences cause the conclusion, "cause" here should be understood as a formal cause.{47} It is in this sense that the elements of a syllogism relate causally. As true, they are different from syllogism in general whose premises may be false. As primary, they require no demonstration, otherwise; they would require demonstration to prove them. These principles capture certain basic truths by being not too broad or narrow (as "appropriate," 72a 6) and so can be identified with immediate principles. As the cause of the conclusion the premises must be better known than the conclusion and be prior to it (72b 17–72a 8). The premises must not only be antecedent in the subject's understanding of the meaning, but also in his knowledge of the fact.

Now the function of a proposition is to predicate a single attribute of a single subject. If either part of this function be taken casually, i.e., merely according to common opinion, then the proposition is dialectical and the resulting argument will be dialectical. If the proposition lays down one part to the exclusion of the other based upon the immediate truth of the accepted part, then the resulting argument will be a demonstration (72a 8–12). The truth of a proposition thus requires a reference to its source.

An important concept which has been employed so far in this explanation is that of an immediate truth. There are two varieties of these immediate truths. The first is the axiom (<u>axiōma</u>) which must be accepted before any discourse whatever may progress, such as the axiom of non-contradiction. No sense can be made of any proposition unless this axiom holds because the essence of "finding meaning" requires a constant, unvarying object which may be examined. The second variety of immediate truth is the thesis (<u>thesin</u>) which is subject to dispute without hindering progress. There are two varieties of theses: definitions and hypotheses. A definition states the essential properties a thing has so that the "what" of a thing is illustrated. In the case of a demonstration this occurs in the first premise through the middle term, which acts as a cause to the predicate extreme. Hypothesis functions in a syllogism through the second premise by affirming the subject's existence or non-existence (72a 15–24). The difference between these two varieties of theses is that the definition requires only that it be understood, while the hypothesis postulates facts, on which the being of the fact inferred depends (76b 35–40, cf. 90b 25–39). Thus, it can be seen that axioms differ from theses in that demonstration must proceed <u>according</u> to the principles of the former and <u>from</u> the principles of the latter.

Another important function of demonstration is to allow knowledge of the reasoned fact as well as knowledge of the fact (APo. 1.13). This distinction bears on the function of the middle term which, as mentioned earlier, must act as a cause in the first premise. For example, to say that "all non-twinkling planets are near" does not establish such a connection. This is because being "non-twinkling" does not cause a planet to be near. It is the other way round, "All planets that are near are non-twinkling." In this second formulation the middle term causes the predicate of the conclusion in the major premise. Thus the knowledge of the fact refers to knowing that the non-twinkling things are near, and the knowledge of the reasoned fact allows one to know that because the planets are near, they do not twinkle. In the former case the connection is accidental, while in the latter case the connection is necessary.

Not only must scientific knowledge yield knowledge of a cause, but the knowledge of the cause must be proximate and not remote. Through a remote cause one could assert that the reason a wall does not breathe is because it is not an animal. The cause "not being an animal" is too remote to count as the proximate cause. It is not in virtue of not being an animal that a wall does not breathe, but because the wall has no lungs. "Having no lungs" is thus seen to be the proximate cause of a wall not breathing. Though there are other restrictions concerning the difference between knowledge of the fact and knowledge of the reasoned fact, these two just described serve to illustrate why knowledge of a fact is accidental and knowledge of the reasoned fact is necessary.

The distinction between knowledge of the fact and knowledge of the reasoned fact illustrates several other useful distinctions. For example, sense perception only allows knowledge of the fact, therefore; there cannot be scientific knowledge through perception alone (APo. 1.31). Opinion also can be distinguished from scientific knowledge because scientific knowledge shows that its object is real and true and cannot be otherwise, while opinion shows what could be true or false and therefore could be otherwise (APO. 1.33). Knowledge asserts a necessary connection between the middle and the extremes, while this connection is not necessary in the case of opinion. The difference between scientific knowledge and opinion comes down to a difference in their objects. This difference makes it impossible for one person to have scientific knowledge and to have opinion about the same object because in the former, the object is necessarily the way it is and could not be otherwise; while in the latter, the object is not necessary and could be otherwise. One cannot think a thing is both necessary and not necessary at the same time and in the same respect. However, while this is true, it is possible for two different people to differ so that one person has scientific knowledge about some x while the other has opinion about x. In this other sense, scientific knowledge and opinion can have the same object. There are, however, two senses of the same object. The first is that a third person could look at some x and judge that that x was the same one that persons a and b gave opinions and had knowledge of respectively. On the other hand, the object is different in that on the one hand it is necessary and on the other it is not. This disallows one person from having knowledge and opinion about the same object. However, with a view to action, it does not matter whether one has true opinion or knowledge if the result is the same (APo. 1.13).

Now one's knowledge of the conclusion of such a demonstration will yield scientific knowledge (APo. 71b 17). So Aristotle seems to follow in the Greek tradition of naming scientific knowledge as the conclusion of a deductive argument (though Aristotle's deductive argument for science, as has been just shown, is much more systematic than that of his predecessors). The key question is how does Aristotle justify the premises for his deduction. Given that the conclusion of a valid deduction follows necessarily from the premises via the rules of logic, the only check on the truth of the argument is the premises themselves.

Aristotle says we have knowledge of the primary premises via induction (APo. 100b 3), which works through intellectual intuition (nous again in a different form). At first glance, Aristotle would appear to be no different from his predecessors who asserted anything they wanted to for a first principle and proceeded from there (uncritical rationalists). Aristotle recognized this as a possible criticism to his theory from those who would claim that everything could be demonstrated from such an account and from those who would claim that nothing could be demonstrated from such an account. The first group's {48} argument rests upon the assumption that the demonstration, which Aristotle characterizes, is circular and so everything can be shown, i.e., one can accept whatever one wants to from the immediate premises and thereby prove anything. The second group feels there is no reason to accept the premises, and so the process could never get off the ground. The question of the plausibility of knowledge in both cases rests upon the acceptance of immediate premises. But how can these premises be accepted without the danger of these two interpretations? The answer must concern itself with how one can know the principles, according to which and from which, demonstration can proceed. The obvious reply is that there must be another form of knowledge which will justify these immediate principles. This type of knowledge turns out to be intellectual intuition.{49}

Intellectual intuition and the process through which it works, induction, enables one to come to a "stopping point." This stopping point confutes those who say that knowledge could never get off the ground, since the basis of their claim is that knowledge of P requires knowledge of Q, and Q would require another entity (proposition?) R and so forth. The infinite regress never allows knowledge to get started.{50}

> They correctly argue that there is an infinite regress since one could not know the latter through the prior if there were no prior (for it is impossible to arrive at an infinite task) . . . (APo. 72b 8–11).

Obviously, one can confute such a regress by apprehending a principle independently somewhere along the way, but this form of knowledge must be different from demonstration for otherwise circularity begins (i.e., legitimating demonstration by another demonstration).

The circularity tact also fails. Aristotle shows that there are very few demonstrations which are reciprocal (viz., only those in the first figure in which the premises are co-extensive).{51} Thus very few demonstrations are validly reciprocal. In fact, of this larger class, the attempts to show circularity merely show "If A then A" (a tautology proving nothing, 72b 35–73a 5). Of the smaller group of syllogisms which are reciprocal, even these demonstrate only in one of their configurations since only one configuration of the premises establishes the <u>dioti</u>. Thus the argument for circularity also fails. The need is still for another mode of knowing besides demonstration and <u>epistēmē</u>.

> We say that not all knowledge is demonstrable but that unmediated is indemonstrable (and clearly this is necessary for if it is necessary to know the prior out of which there is demonstration, then it stops at the immediates—these are necessarily indemonstrable). (APo. 72b 18–22).{52}

Thus Aristotle must affirm knowledge of immediates. The mode of this is induction <u>and</u> the power of the mind which operates is <u>nous</u>. Since <u>epistēmē</u> is also a power of the mind which works <u>through</u> the <u>process</u> of demonstration, it is <u>properly</u> contrasted with <u>nous</u>: "not only do we say <u>there</u> is <u>epistēmē</u>, but that <u>there is</u> a principle of <u>epistēmē</u> by <u>which we know</u>

definitions" (72b 23-25).{53} Later in B 19 Aristotle discusses the process which yields these unmediated principles.{54} This process can be reconstructed as follows:

1. Sense perception of the particular. (99b 35).

2. A developed power of retention from the persistence of sense perceptions. (99b 36-100a 1).

3. This power of retention is memory. (100a 3-4).

4. The addition of multiple memories in conjunction with continued sense input creates experience, empeiria. (100a 5).

5. From experience (i.e., being able to identify the common element in the memories and to match it against sense input) arises a universal through induction. (100a 5-8).

6. Induction admits the universal into the soul by the power of nous (in the case of unqualified knowledge) and by the power of doxa, opinion (in the case of general knowledge, gnōsis). (100b 5-100b 9).

There are several different interpretations of this difficult text which commentators have come up with,{55} but they need not concern us directly for our purposes in this treatise.{56} What is clear is that induction, epagōgē, operates via aisthēsis, mnēmē (memory), empeiros (experience) and nous. When a particular group of common facts are present in the mind, then intellectual intuition arranges these common elements into a universal. It is not clear just how many cases are necessary for an induction. In APr. 68b 26-30, it seems as if all the cases must be considered while in Topics 105a 13-16 one or two cases seem to be enough. Perhaps it would be fair to conclude from these two passages that one must go through some appropriate number of cases when practically employing induction since complete enumeration of all the cases is rarely possible. What number is "appropriate"? This would have to depend upon the practitioner of induction, but it is clear that too few might lead to gross error, while too many might unduly impede progress.

These two forms of induction are called "imperfect" and "perfect." The latter examines every case and is only practically operative when one is examining species, while the former only examines an appropriate number and is applicable to individuals (see my discussion in Chapter Five). Thus both forms work together. To form a species one uses imperfect induction, and to use specific knowledge to form genera one employs perfect induction. The actual apprehension of the universal is an intellectual process which was first occasioned by an act of sensation. It would be incorrect to demand that there be no cognitive act in an empirical investigation (an extreme claim even for an uncritical empiricist). Such a demand would, according to Aristotle, only give rise to recognition of the fact, <u>hoti</u> (APo. 1.31). Likewise, knowledge of only the reasoned fact is tied to no actual phenomena (as the uncritical rationalists would have it). These two sides thus correspond in a rough way to the principal strains of pre-Aristotelian natural philosophical theory.

The criterion for this selection is the essential attribute which all of these memories share. Not just any commonly shared trait will do, for many of these may be flimsy and accidental, but rather it is only the common essential attributes which will count in forming a universal. When multiple memories are set together the common features are enumerated so that a single experience might be fashioned which contains the essential attribute(s) which all these images share. "It is from memory that men acquire experience, because the numerous memories of the same thing eventually produce the effect of a single experience" (Metaph. 980b 26-981a 1, cf. APo. 100a 5). This universal embodies those attributes which make that event or thing exactly as it is and not otherwise. Each particular memory participates in this newly created universal in virtue of the commonly shared essential attribute(s). Thus it can be seen that <u>nous</u> is not capricious in selecting certain common traits over others in generating a universal--only essential attributes will do because only these truly define a thing. It will be this more technical sense of 'empirical' that will stand for 'critical empiricism' in this book.

Scientific knowledge requires both <u>hoti</u> and <u>dioti</u> APo. 1.12). Thus any definition of critical empiricism which excludes this new conjunction between <u>aisthēsis</u> and <u>nous</u> cannot satisfy the rigorous criteria established here by Aristotle.

There are other meanings of induction which also must be analyzed. In addition to working from individuals to species (imperfect induction), induction can work from species to some larger group (perfect induction). In <u>Rhetoric</u> 2.23, three of our cases cited use this second form of induction (cf. APr. 68b

20). Of course the danger that could occur here is that the species may not have been established by induction. For the species is created from individuals. There are no species walking about, only members of a species. If one takes a species as given without proceeding first through imperfect induction from individuals, then said species may be in error and any further inductions from general to more general (species to genus for example) would only compound the original error. Indeed, perfect induction is only properly executed when the general groups it works with have been formed from imperfect induction. Remember that imperfect induction is "imperfect" because it does not consider all cases, but only an "appropriate" number. Thus perfect induction depends to a large degree upon the groups given to it by imperfect induction. Both forms of induction are necessary for practicing biology, however; it is only perfect induction which is useful in creating biological theory. This is because when one creates a theory one seeks a certain degree of generality which can only be found in discussing groups. The biologist does not discuss Rover, Fido and Spot, but rather is interested in "dogs." When analyzing a cell he is interested in the behavior of mitochondria in general and bases his work upon his understanding of these groups. Ideally, for Aristotle, the principal work of the biologist is in working from these established groups via perfect induction to generalizations about their respective functions. However, this description applies to a mature science. As biology is just beginning with Aristotle, he must spend a great deal of time creating the groups through imperfect induction. Much of this groundwork occurs in the Historia animalium, but it occurs in other places as well. After the work of imperfect and perfect induction, then one has finished the discovery phase of a science (APo. 2) and one embarks upon the explanatory axiomatizing phase (APo. 1). This latter phase occurs in a finished science which is being presented in its logical order. When one engages in this exercise one is concerned with making the formal presentation as tight as possible according to the rules laid down in APo. 1 (discussed earlier). Aristotle never got to this point in biology or in any other science so that we never see such an exposition in his writing, but one can look at some of the work of the logical empiricists in our own century to see people engaged in a similar variety of formal, theoretical exposition.{57} Despite the fact that Aristotle never reaches this stage, he still may be said to carry out the principles of discovery outlined in APo. 2, in a general way, in both P.A. 1 and in his actual biological investigations.

There are a few further complications when one considers the data upon which imperfect induction is based. The *phainomena* for imperfect induction must arise from *aisthesis* (*G.A.* 756a 5, 760b 27 ff.). The alternative to this would be to focus upon those problems, *aporiai*, and common opinion, *endoxa*, which Wasserstein and Owen have documented. This, of course, would lead to the uncritical rationalism discussed earlier. Certain commentators have asserted that this is exactly what happens in Aristotle's actual scientific investigations.{58} This occurs when *epagōgē* begins dialectical arguments (*Topics* 105a 10-19). This is sometimes called "dialectical induction" in which common opinion begins dialectic. Since induction normally begins with the phenomena and moves to the universal and since in this case induction begins with common opinion (in dialectic--what is accepted by all or most men or by the wise, *Topics* 100b 21-23), then in this instance *phainomena* must be *endoxa*. Let this usage of phenomena be termed phenomena-2 and our former usage phenomena-1. As mentioned earlier, this difference is not absolute, but rather of degrees of emphasis. It is in this way that dialectic can check first principles of science. By raising searching difficulties on both sides of a subject, dialectic is one way that false principles can be rejected (*Topics* 101a 35-101b 4). Thus dialectic acts as a method of negative verification. By rejecting the false principles, it is assumed that only the true ones remain.

In this case, the "facts" (*phainomena*) are common opinion which are, of course, once removed from the thing in question (unless the thing in question is that very opinion, e.g., through an opinion poll about something). But when the phenomena become common opinion, then the *aporiai* become logical or philosophical puzzles. The reason for this is that an unexplained fact or difficult case in the realm of common opinion is a paradox (*paradoxos*). Being contrary to what is generally held, it becomes a puzzle to be explained. Zeno's paradoxes discussed in the *Physics* play just such a role. They are logical puzzles which Aristotle feels he must explain. He does this by laying down principles and following them to their logical conclusion. The key to judging whether these principles are empirical or not is whether they are laid down from phenomena-2 or phenomena-1. If they arise from considerations aimed solely at solving logical paradoxes, then they cannot be critical empirical theories. A possible exception to this would be the body of scientific data which might comprise the *endoxa*. This was discussed earlier, in a different case. The key to this other sense of phenomena is that the *explanans* be tied to the *explanandum* through firsthand or secondhand observations and that, according to rigorous methodological standards, nous and aisthēsis are welded together in a system of checks and balances. In saying this, the role of reason cannot be underplayed. As a matter of fact, scientists always seem to be interested in solving problems that arise in existing theories by positing new hypotheses and new theories.{59} Many times they will have no empirical basis for adopting certain first principles. Copernicus, for example, chose the heliocentric hypothesis for an archē (first principle) because it was more consistent with the Aristotelian doctrine of perfect circular motion (in that he could eliminate the equant).{60} It also accorded with the Platonic/neo-Platonic penchant for circularity (cf. Timaeus 34b, 36c). For Copernicus it was important that he move away from the Ptolemaic System and its theoretical inconsistency, viz., the equant's non-uniform motion. The only way he felt he could do this was to postulate another first principle which would solve this problem. His motivation for positing the heliocentric hypothesis was solving a problem of putting astronomy in accord with Platonic influences and Aristotelian physics.{61} These are "rationally oriented" arguments, but that does not mean that Copernicus was an uncritical rationalist or that he was wrong in using them. Sometimes empirical verification must come long after a theory has been offered.{62}

This may be more characteristic of physical than biological theories perhaps because the crucial aspects of a biological explanation depend upon very close observation. One may create a rather sophisticated theory about physics without having made any observations beyond common experience. There are numerous exceptions to this, of course, such as microscopic events and rather intricate physical events, but in a rough and ready sense there are a vast number of physical "facts" which we all know and could describe without controversy. All people know that when you drop a stick, it will fall (though perhaps some of the details such as how the stick will fall relative to a heavy stone may need to be observed closely). The critical aspect of explaining this is not knowing that the stick falls, but why it falls. The hoti is taken as given; all of us experience the same events. But with biology the case is different. Arriving at the 'fact' is no easy matter. How many chambers does a heart have? What does blood do? How do the lungs operate? Such questions as these are not readily available to anyone unless he dissects an animal. It seems obvious that the answers to these questions need to be prior to the explanations concerned with why. The fact must precede the reasoned fact. Before one can explain why a whale has a blowhole, one must first know that a whale has a blowhole. This is why it is necessary to derive one's biological account from actual organisms. In simple examples from physics, such as a stick falling, this emphasis is perhaps unnecessary since the principal dynamics are rational deliberations about why. But in biology this further step is also required in a complete account of discovery.

There is one final distinction that must be made. Some principles may be posited rationally, but only because the theory which they create takes time to establish its empirical verification. Einstein could not offer empirical verification for his theory of general relativity when he offered it, but he felt that in principle it was possible. Such cases must be distinguished from those in which no such verification is ever sought or expected, such as the theories mentioned in 1.1. These latter theories must fall into a special class of being truly a priori (uncritical rationalism), while the former are in a category which, while certainly emphasizing the rational at present, portend future empirical support. And though the genesis of such a theory may not have been in aisthēsis and nous as an integral pair, the theorist does not deny that the first principles (and the corresponding theory) may be ultimately confirmed or denied by such data. That these theories commit themselves ultimately to being just as dependent upon sensory data in the same way as other critical empirical theories do is evident from their challenge of future empirical verification. Because of this these latter theories will also be termed "critically empirical."

Thus the question of the relative places of nous and aisthēsis has to be answered tenatively. There are no hard and fast lines of demarcation between critical empiricism and uncritical rationalism and uncritical empiricism. Rather the distinctions are subtle, and especially in mediate areas, one can only point to degrees of emphasis on a continuous scale. However, it is clear that there are certain real differences in the method of theory construction in these various modes which are significant to the end product achieved. Aristotle seems to have solidified a methodological trend among the Coan writers as well as among Empedocles and Anaxagoras (perhaps the Atomists also?) toward critical empiricism where, through a detailed theory of induction, the relative places of these various elements are set down in a precise manner.

Following from this general background and outline of Aristotle's general philosophy of science, we will now turn our attention to the specific methodology created for his biological investigations, and finally (in Chapter Five) we will note how the two interrelate.

Notes to Chapter One

1

Interesting studies on the development of these words in the context of the development of ancient Greek thought has been done by J.H. Lesher, "Perceiving and Knowing in the Iliad and Odyssey" *Phronesis* 26.1 (1981): 2-25; Bruno Snell, "Die Auesdrucke *fuer den* Begriff des Wissens in der Vorplatonischen Philosophie" *Philogische Untersuchen* 29 (Berlin, 1924), pp. 24 ff.; and Kurt *von Fritz*, "*Noos*, *Noein* in the Homeric Poems," *Classical Philology* 38 (1943): 79-93; "*Nous*, *Noein* and their *Derivatives* in Pre-Socratic Philosophy" *Classical Philology* 40 (1945): 223-242 & 41 (1946): 12-34; "Der *Nous* des Anaxagoras" *Archiv fuer Begriffsgeschichte* 9 (1964): 87-102. I side with *Lesher* on the "*nous* /*gignoskein*" dilemma, but accecpt their common conclusion on the emergence of the bifurcation distinction. The major bone of contention here is *when* such a distinction occurs and when this is developed. This *is still* an open issue and my comments are only general reflections on this trend. A specific determination of this will require the energies of an entire project.

2

Sometimes this recognition comes from sight itself, i.e., some visible qualities in the stranger as in *Iliad* 3.386 ff. At other times the recognition occurs without any reference to visible qualities (*Odyssey* 1.322, 4.653 etc.).

3

Lesher does want to make a case for this possibility occurring as early as the *Odyssey*. Whereas I agree with Lesher about the possibility of oral philosophical poetry, the lack of sustained argumentation (an indication of the bifurcation of reason and sense experience) seems to warrant a later date for "critical philosophy."

4
The dates here are rather inexact due to difficulties in determining births and deaths for these philosophers. In lieu of that I have tried to give a rough guide to the "floruit" of the individual concerned.

5
For two interesting treatments of this post Aristotelian school see: Ludwig Edelstein, "Empirie und Skepsis in der Lehre der griechischen Empirikerschule," Quellen und Studien zur Geschichte der Naturwissenschaften und der Medizin, Band 3, Heft 4 (Berlin, 1933), pp. 45-54; and Karl Deichgraeber, Die griechische Empirikerschule (Berlin, 1930).

6
For a discussion of this point of view see: G.H. Lewes, Aristotle: A Chapter from the History of Science (London, 1864), p. 49ff. and Wolfgang Kullmann, Wissenschaft und Method (Berlin, 1974), p. 88.

7
Ingemar Duering, "Aristotle's Method in Biology," Aristote et les Problèms de Méthode (Louvain, 1961), pp. 213-221.

8
G. Pouchet, La Biologie Aristotelique (Paris, 1885), p. 19; cf. Maurice Manquat, Aristote Naturaliste (Paris, 1932), pp. 31-50.

9
This view is taken by Hermann Fraenkel, "Xenophanesstudien," Hermes 60 (1925): 174-192.

10
References made to the Pre-Socratics are made according to Hermann Diels, Die Fragmente der Vorsokratiker 6th ed., revised by W. Kranz, 3 vols. (Berlin, 1952).

11
Some like von Fritz, believe that the two have not been split (loc. cit.), whereas others do believe this has happened and try and draw conclusions from the odd results which follow. Much of this has involved interpreting the manner of Heraclitus' utterance. Is it gnomic and full of riddles? What do these portend? Analyses of this variety has been taken on by C.J. Emlyn-Jones, "Heraclitus and the Identity of Opposites,"Phronesis 21 (1976): 89-114; Uvo Hoelscher,

Anfaengliches Fragen: Studien zur fruehen grichischen Philosophie (Goettingen, 1968), pp.136-42, 144-48; and U. Moulines, "Un problema de interpretacion del logos en Heraclito," Boletin del inst. de Estudios Helenicos 6.2 (1972): 77-91.

12 B40-πολυμαθίν νόον ἔχειν οὐ διδάσκει· 'Ησίδον γὰρ ἄν ἐδιδαξε καὶ Πυθαγόρην αὖτίς τε Ζενοφάνεά τε καὶ 'Εκατοῖον. 104. τίς γὰρ αὐτῶν νόος ἤ φρήν; δήμων ἀοιδοῖσι πείθονται καὶ διδασκάλωι . . .
Bruno Snell, op.cit., pp. 72ff., makes an interesting case uniting these two by trying to show that they point to knowledge acquisition via the object rather than the subject.

13 ξὺν νόῳ λέγοντας ἰσχθρίξεσθαι χρὴ τῷ ξθνῷ πάντων, ὄκωσπερ νόμῳ πόλις, καὶ πολυ ἰσχυροτέρος. τρέφονται γὰρ παντες οἱ ἀνθρώπειοι νόμοι ὑπο ἑνὸς τοῦ θείου· κρατεῖ γὰρ τοσοῦτον ὀκόσον ἐθέλει καὶ ἐξαρκεῖ πᾶσι καὶ περιγίωεται.

14 Logos here is perhaps the hidden spirit of reason, or perhaps it stands for some divine element in understanding. At any rate it is that which gives meaning to some account.

15 This analysis is made by R.A. Prier, "Symbol and Structure in Heraclitus" Apeiron 7.2 (1973): 23-37.

16 The idea here is that in an analogous way to the characters in the Iliad and Odyssey seeing beyond the appearances of the disguised gods, the reader of these fragments sees beyond the partial truths of seeming paradoxes to find the true intended meaning; cf. C.J. Emlyn-Jones and Uvo Hoelscher.

17 These words--nous, noein, noema or noesis--occur in the first half at: B2, l. 2, 8; B4, l. 1; B6, l. 1, 6; B7, l. 2; B8, l. 8, 34, 36, 50. It further should be noted that Jonathan Barnes, The Presocratic Philosophers (London, 1979), pp. 155-160, makes Parmenides the first philosopher to break off on a new direction; see p. 158, n. 6 on 'nous.'

18
There has been a great deal of secondary literature written on this. See especially: Karl Reinhardt, Parmenides und die Geschichte der griechischen Philosophie (Bonn, 1916); Hans Schwabl, "Sein und Doxa bei Parmenides," Wiener Studien, 70 (1957): 278-89; Karl Diechgraeber, Parmenides Auffahrt zur Gottin des Rechts: Untersuchungen zum Prooimion seines Lehrgedichts, Akademie der Wissenschaften und der Literatur in Mainz: Abhandlunger der Geistes—und Sozial-wissenschaftlichen Klasse, Jargang 1958, 11 (Wiesbaden, 1959), pp. 629-724; Jaap Mansfeld, Die Offenbarung des Parmenides und die menchliche Welt (Assen, 1964), ch. 3. These writers deal with the problems of the two halves in their various contrasts and similarities.

19
Two works that especially show this are: W.J. Verdenius, "Der Logosbegriff bie Heraklit und Parmenides, II" Phronesis 12 (1967): 99-117; and Alexander P.D. Mourelatos, The Route of Parmenides: A Study of Word, Image and Argument in the Fragments (New Haven, Conn., 1970), pp. 222-63.

20
A. Wasserstein, "Greek Scientific Thought" Proceedings of the Cambridge Philological Society n.s. 8 (1962): 51-63.

21
ibid.

22
Wasserstein quotes Proclus, "In their desire to demonstrate the uniformity of heavenly motion the astronomers have only, without noticing it, succeeded in demonstrating that these movements are essentially irregular . . . (Hypotheses, Halma, pp. 150-151).

23
Wasserstein, p. 56, "It is clear from Proclus' remarks that he is willing to regard not only circularity but even uniformity as expendable assumptions." G.E.R. Lloyd makes a similar point about Greek Science in his book Magic, Reason and Experience (Cambridge, 1979), ch. 3.

24
In Aristarchus' work, On the Sizes and Distances of the Sun and Moon, he introduces figures for the apparent purpose of showing a method. His figures are wildly false. But in Arenarius of Archimedes there is the account that Aristarchus

"found that the sun's apparent size is about 1/720th the part of the Zodiac circle," and that equals 1/2 degree indicating that Aristarchus knew the correct figures.

25
This argument has been made in connection with the Greek word 'phainomena' by G.E.L. Owen, "TITHENAI TA PHAINOMENA" Aristote et les Problèms de Méthod (Louvain, 1961), pp. 83-103.

26
There is not space enough here to enter this great dispute. My reading of Empedocles accords with the work done by Jean Bollack, Empédocle: Introduction à l'Ancienne Physique vol. 1 (Paris, 1965); vols. 2-3, Les Origines (Paris, 1969). Bollack shows the tradition dilemma to be a pseudo-problem. Cf. Uvo Hoelscher, "Weltzeiten und Lebenszyklus," Hermes 93 (1965): 7-33; and Friedrich Solmsen, "Love and Strife in Empedocles' Cosmology," Phronesis 10 (1965): 109-148. One can compare this to the traditional view that there is an alternation between the cycles: J.E. Raven, The Pre-Socratic Philosophers (Cambridge, 1957) and W.K.C. Gutherie, A History of Greek Philosophy (Cambridge, 1965) and most especially the very fine work by D. O'Brien, Empedocles' Cosmic Cycle: A Reconstruction from the Fragments and Secondary Sources (Cambridge, 1969). O'Brien's book also contains an excellent bibliography.

27
This problem has been stated by Zeller that the religious teachings "stand in no visible connection with the scientific principles [of his physics]." Die Philosophie der Grechen vol. 1, 5th ed. (Leipzig, 1892), p. 806. Jaeger makes a similar point in Paideia, English tr., 2nd ed., vol. 1 (N.Y., 1945); cf. G. Vlastos who says, "it would be futile to try to supply it [an explanation which would put the parts together] by rationalizing the theology of the mystic or mystifying the logic of the cosmologist" Philosophical Quarterly 2 (1952): 121. More recently this incompatibility thesis has been stated by A.A. Long, "Thinking and Sense-Perception in Empedocles: Mysticism or Materialism?" Classical Quarterly, n.s. 16 (1966): 255-276.

28
Fr. 21-ὑπ' ἀφαυρότητος αὐτῶν (sc. τῶω αἰσθήσεων) οὐ δυνατοί ἐσμεν κρίωειν τἀληθές. 21a-ὄφις γαρ τῶν ἀδήλων τὰ φαινόμενα.

29
If "phenomena" is taken in Wasserstein's sense, then the "putting together" would be the answering of some problem which is then deductively carried through to its conclusion.

30
It is interesting to note that there are no "B" passages in Diels-Kranz in which Anaxagoras himself refers to these seeds as "homoiomerous." If he did not, then part of this dilemma might be immediately solved. The references to the homoiomerous (as such) occur at A4 (Ar. *Phys* 187a 23), A3 (Ar. *De Caelo* 302a 28), and A 46 (Aetius, 1,3,5). However even if this were accepted the passage of B4 must be accounted for.

31
B4 . . . καὶ σπερμάτων ἀπείρων πληθος οὐδὲν ἐοικότων ἀλλήλοις. Οὐδὲ γὰρ τῶν ἄλλων οὐδὲν ἔοικε τὸ ἕτερον τῷ ἑτέρῳ. τούτων δὲ οὕτως ἐχόντων ἐν τῷ σύμαντι χρη δοκεῖν ἐνεῖναι πάντα χρήματα.

32
The problem has been clearly set out by Cornford: "Anaxagoras' theory of matter . . . rests on two principles which seem flatly to contradict one another. One is the principle of Homoeoerity: A natural substance, such as a piece of gold, consists solely of parts which are like the whole and like one another . . . . The other is: 'There is a portion of everything in everything,' understood to mean that a piece of gold (or any other substance), so far from containing nothing but gold, contains portions of every other substance in the world. Unless Anaxagoras was extremely muddleheaded, he cannot have propounded a theory which simply consists of this contradiction." F.M. Cornford, "Anaxagoras' Theory of Matter," *Classical Quarterly* 24 (1930): 14-30, 16. Trying to fit these two parts together has been the task of numerous commentators. Among the most important discussions have been: A.L. Peck, "Anaxagoras and the Parts," *Classical Quarterly* 20 (1926): 57-62; "Anaxagoras: Prediction as a Problem in Physics," *Classical Quarterly* 25 (1931): 27-37, 112-120; Cyril Bailey, *The Greek Atomists and Epicurus* (Oxford, 1928), Appendix 1; Gregory Vlastos, "The Physical Theory of Anaxagoras," *Philosophical Review* 59 (1950): 31-57; J.E. Raven, "The Basis of Anaxagoras' Cosmology," *Classical Quarterly* n.s. 4 (1954): 123-37; Charles Mugler, "Le Problème d'Anaxagore," *Revue des etudes grecques* F.M. Cornford's Interpretation, *Classical Quarterly* n.s. 8(1958): 67-81; Colin Strang, "The Physical Theory of Anaxagoras," *Archiv fuer Geschichte der Philosophie* 45

(1963): 101-118; Michael Stokes, "On Anaxagoras," *Archiv fuer Geschichte der Philosophie* 47 (1965): 1-19, 217-50; W. Schwabe, "Welches sind die materiellen Elemente bei Anaxagoras," *Phronesis* 20 (1975): 1-10.

33
According to DK A3 (Ar. Metaph. 984a 11), Anaxagoras was older than Empedocles, but came to his philosophical writing after him.

34
An interesting study of this can be found in Ludwig Edelstein, "Greek Medicine in its Relation to Religion and Magic" *Bulletin of the Institute of the History of Medicine* 5 (1937): 201-246; and in Lloyd, op.cit. ch. 1.

35
Whether there is such a distinction in "schools" is a much debated question. Littré seemed to think there were six Cnidian treatises: *Epidemics* 2, 3, 4; *The Nature of Woman*, *On Seed* and *The Nature of the Child* (Littré, 7, 304-310). Ilberg in "Die medizinische Schrift ueber die Siebensahl und die Schule von Knidos," *Festschrift fuer Lipsius* (Berlin, 1894), pp. 33ff., "Die Aerzteschule von Knidos," *Berichte Saechs. Akad. Wissenschaft* (Leipzig, 1925), p. 76, raises this number to twelve. Edelstein in *PERI AERON und die Sammlung der hippokratischen Schriften* (Berlin, 1931), p. 159, accepts only *Regimen in Acute Diseases* 1, 2; *Epidemics* 2 and *Internal Affections*; while L. Bourgey, *Observation et Experiénce chez les Médicins de la Collection Hippocratique* (Paris, 1953), pp. 50-53, sees only *Internal Affections* and *Epidemics* 2 as truly Cnidian. But the criteria for this attribution seems a bit shaky. Littré saw the gynocological works as being a basis for differentiating the Cnidians from the Coans (cf. 1, 440-464), while R. Joly looks to the various versions of the humours to give the answer, *Hippocrate* vol. 11 (Paris, 1970), p. 27. I.M. Lonie tries to show how a system of unique aetiology, symptomatoloty and therapy mark the Cnidians as distinct from the Coans, "The Cnidian Treatises of the *Corpus Hippocraticum*" *Classical Quarterly* n.s. 15 (1965), pp. 1-30. This view has been challenged by Wesley Smith, "Galen on Coans versus Cnidians," *Bulletin of the History of Medicin* 47 (1973), pp. 569-585. Smith claims that no such distinction can be made based upon the evidence we possess. This 'inference from the evidence' argument is elaborated by G.E.R. Lloyd, "The Hippocratic Question" *Classical Quarterly* n.s. 25 (1975): 171-192, as he tries to discredit the traditional techniques

which attribute texts to one particular writer. This is indeed a difficult question, however; it does seem clear to me that certain groups of treatises do form natural groups. This is true of the gynocological writings which have certain characteristics. These characteristics are shared in Epidemics 2, 3 and possibly other writings. If one does not want to call them Cnidian, any other name will do, the name is not what is at issue, but rather methods of doing medicine and constructing theories.

36
Kuehn, 10,6.

37
This pragmaticism contrasted with the school of Cos. The detailed rules of the Cnidians were meant to be followed "au pied de la lettre" irrespective of the individual constitution of the parient which is so regarded by the Coan school. For an elaboration of this argument see: K. Sudhoff, Kos und Knidos (Munich, 1927), p. 97.

38
Edelstein makes a similar point in "The Distinctive Hellenism of Greek Medicine" Bulletin of the History of Medicine, 40 (1966): 197-255, part 1.

39
Ibid.

40
Cf. Owen, op.cit.

41
I assume that these other natural philosophers are Empedocles and possibly Anaxagoras. The guess follows both from the chronology and from the nature of the attack.

42
For the following Hippocratic quotations I will be following W.H.S. Jones's Loeb translation. Ancient Medicine.

43
Polybus, the son-in-law of Hippocrates, is attributed with The Nature of Man by Aristotle.

44 The thawing water is lost through evaporation.

45 Dr. E.T. Withington, "Introduction," Hippocrates Vol. 3 (Cambridge, Mass., 1928).

46 This requirement may not apply to Aristotle's biological investigations since they concern natural and not mathematical sciences. I will maintain later that these two are different, thus Aristotle's requirement here may be superfluous to his biological science.

47 Compare to the other causes such as the final cause, APo. 2,11.

48 I have switched Aristotle's order for clarity in my presentation.

49 APo. 72b 5-18.

50 Cf. other infinity passages: Bonitz 74b 41-47; Metaph. 99b 16-27; APo. 82b 29, 83b 6, 84a 3; Caelo 272a 3; Phys. 238a 33. For an interesting rendition of this argument see Jonathan Barnes' note to this passage in his Aristotle's Posterior Analytics (Oxford, 1975).

51 The mechanics of this tricky passage are best set out by W.D. Ross, Aristotle's Prior and Posterior Analytics (Oxford, 1949), p. 513.

52 Philoponus in his commentary [ed. Wallies (Berlin, 1909)] makes an interesting argument for the unmediated premises being grasped by nous. First, it is known that the conclusion of the demonstration is known via the middle, which itself never occurs in the conclusion. The middle therefore cannot be demonstrated. If it is to be known (and it must be known according to the rules of demonstration), then the mode of its being known must be nous. Therefore the middle, along with its connection to the extremes, must be known immediately and not demonstrably.

53
Two points: a. Is nous the principle of episteme? Both Philoponus and Themistius [Analyticorum Posteriorum Paraphrasis ed. Wallies (Berlin, 1900)], agree that it is—saying that it provides the underpinnings for discourse. Barnes complicates the passage saying that what "is referred to as understanding at 72b 18 becomes a principle of understanding at 72b 24. In fact, the difference between 72b 18 and 72b 24 merely marks two different ways of expressing the one fact urged in 72b 18–24" (p. 109). Barnes seems to be following Zarabella, In duos Aristotelis libros Posteriorum Analyticorum Commentaria (Venice, 1582). It is true that Aristotle does not keep to a technical use for his terms, but it seems clear to me that there are two types of knowledge with two different objects. b. Are these objects definitions only? There are several passages which suggest this: 90b 27, Top. 158a 33; De An. 402b 16–26; Metaph. 998b 5, 1034a 30–2, 1078b 23. However, it would seem that axioms and the other brand of thesis, hypothesis, would fall under this category, too.

54
For earlier references to this chapter see: 71b 16, 72b 18–25.

55
A good discussion of the interpretations of the Greek and Latin commentaries occurs in J.M. Le Blond, Logique et Methode chez Aristote (Paris, 1939), pp. 129–141.

56
Two interesting treatments of this passage and related subjects can be found in Robin Smith's, "The Indemonstrability of Principles in Aristotelian Science" presented to the Eastern Division APA, December, 1981; and Deborah Modrak's "Imagination and First Principles" presented to the Western Division APA, April, 1982.

57
The relationship between observational statements and theoretical statements has been analyzed from this point of view by: Frank Ramsey, "Theories," in The Foundations of Mathematics and Other Logical Essays ed. Richard B. Braithwaite (London, 1931); Rudolf Carnap, "The Methodological Character of Theoretical Concepts" in Herbert Feigl et al. eds. Minnesota Studies in the Philosophy of Science vol. 1 (Minneapolis, 1956), pp. 38–76, "Beobachtungssprache und theoretische Sprache" Dialectica 12 (1958-9): 238–248; and Çarl Hempel, Fundamentals

of *Concept Formation in Empirical Science* (Chicago, 1952).

58
Cf. Pouchet and Manquat (n.8); Duering (1961), p. 220 and *Aristotle's De Partibus Animalium and Literary Commentaries* (Goteborg, 1943), pp. 36-37.

59
Albert Einstein says in a forward to Stillman Drake's translation of Galileo's *Dialogue Concerning the Two Chief World Systems* (Berkeley, 1967), pp. xvi-xix, "There is no empirical method without speculative concepts and systems." And in applying this to Galileo, whose theories were ahead of techniques for verification, Einstein continues, "Moreover, the experimental methods at Galileo's disposal were so imperfect that only the boldest speculation would possibly bridge the gaps between the empirical data."

60
For a further discussion of the equant which was formed by making the center of an epicycle "off center" so that the line from the new point to the edge of the circle was no longer the radius, but a line segment whose end point on the edge of the circle would appear to speed up and slow down, thus accounting for the various speeds observed in retrogression, see: Thomas S. Kuhn, *The Copernican Revolution* (Cambridge, Mass., 1957), pp. 70-71.

61
Nicolaus Copernicus, *De Revolutionibus, Occasional Notes of the Royal Astronomical Society*, 10 (May, 1947), Bk. 1, 4 & 5, pp. 9-10, cf. ch. 8.

62
Before a new theory is posited it may be impossible to empirically break out of the hold that the old theory has on us. This argument concerning conceptual frames of reference influencing our perception is made by Norwood Russell Hanson, *Patterns of Discovery* (Cambridge, 1958), pp. 10-30.

THE CONTEXT

CHAPTER 2

THE METHOD

2.1 PHILOSOPHY OF BIOLOGY AND THE BIOLOGIST

As mentioned earlier, P.A. 1 is the most important repository of methodological theory for Aristotle's biology. Before examining the two key opening questions, it is of some interest to note the opening passage of the book which is frequently overlooked. Aristotle says,

> In all study and investigation, be it humble or noble, there appear to be two types of proficiency. One is scientific knowledge of the subject; the other is an educated person's competence. For the latter has the power to judge what is put forth properly and what is not. (P.A. 639a 1-7).

Here Aristotle distinguishes between the specialist (the practitioner who has *epistēmē*) and the generalist who has a sort of educatedness (*paideia* 639a 4). The specialist is one who knows truths of one particular discipline but not necessarily those of other disciplines. This, of course, is contrary to the Platonic notion that the man who has *epistēmē* must have knowledge of all subjects. Aristotle is asserting here that a man may be an expert at biology without necessarily having knowledge of another discipline, such as mathematics or astronomy. But what would this sort of knowledge be? It has to be concerned with biology alone, therefore; it must concern itself with that which is uniquely biological: the phenomena of living things. The expert in biology (here Aristotle is really concerned with zoology){1} has a command of the appearances relevant to the animal realm or some part of the animal realm. Beekeepers, for example, may know a considerable amount about bees, and fishermen may be experts on fish. Since these people spend a great deal of time with these animals, it is natural that they should know the facts about such. This does not necessarily mean that the expert understands the principles or causes of these facts. In fact, as David Balme has pointed out, these two classes are sharply distinguished by Aristotle as: a) experts who only know the fact (the *hoti* discussed in the last chapter), and b) those who know the facts and the causes (639a 8-11).{2}

Distinct from both of these categories is the generalist who has knowledge of all (or most of) the disciplines. This person might be characterized as the philosopher who not only speculates about causes in one discipline, but does so about all or many of the various sciences. One may view this role in light of the discussion in Chapter One on critical and uncritical rationalists and empiricists. It is evident that the philosopher, who speculates on causes which affect several disciplines without being an expert in one or more of these areas, may tend towards the uncritical rationalist pole. The extent to which he does so depends upon the extent to which he also makes use of experts (either of variety [a] or [b]). To be thoroughly in the camp of critical empiricism it is requisite not only to speculate about causes, but also to be grounded personally, to some extent, in the data themselves.

Thus, there are two functions being affirmed for the philosopher of biology: 1) empirical investigations, and 2) speculations on the causes or study (*methodos*) of the science *per se* or of *that* science in *relation to others*. These two *modes* represent *the two* foci of this book: *the method* and the practice. This chapter and Chapter Three will discuss P.A. 1 while Chapter Four will briefly examine certain applications in the practice.

It is important to keep these two modes separated and to realize when Aristotle is engaged in one as opposed to the other. These two activities are separate according to Aristotle in this passage. This is an important consideration (one which is often overlooked by the fact that Aristotle was both a theoretician and a practitioner of biology). It would be unfair to criticize Aristotle for not doing biology when he is engaged in method and *vice versa*.

Specifically, in this chapter, I plan to show first Aristotle's principal interest in engaging in systematics and not taxonomy itself. Next I will tackle the problem of *genos* and *eidos* and their meanings in the biological works. This enables me to examine the status of Aristotle's system of classification in light of modern systematics. Finally, this latter theme is examined more thoroughly through the specific method of classification via Aristotle's discussion of dichotomy. In the end, it is critical empiricism which seems to be Aristotle's ultimate motivation in his arguments against division and other uncritical rational modes of classification.

2.2 OPENING QUESTIONS

The first key question to De partibus animalium occurs at 639a 15.

> I mean, for example, should one take each single species and state its differentia independently, for example Man's nature or Lion's nature or Ox's nature and so on, or should one first posit the attributes which they have in common because of a common character which they share. (P.A. 639a 15-19).

Two alternatives are suggested. One would have the biologist examine each species and catalogue it by itself: that is, give a physical description which is unique, e.g., length, height, width, weight and shape. In other words, each species is described with reference to itself only. There would be no interrelations made between various descriptions (until perhaps the initial classification was over). This would be something like the Cnidian procedure of medical treatments which were discrete and unrelated. According to Aristotle's account of definition, this would be impossible. A part cannot be known except in the context of increasing generality. Beyond the necessary requirements of genus and species, each primary substance also can be fit into an appropriate category along with its properties (cf. Categories 4).

The second suggestion deals with establishing common attributes. Now at first sight this appears to be a criterion for a system of biological classification. There is some support for this (to be discussed later), but besides this implication there is another rationale for proceeding in this way. If one listed each species separately, then there would be quite a large number of discrete cases, each of which would require a separate explanation. These explanations, like their descriptions, would not be related. Aristotle rejects this as unwieldly and vague.

> So that if we describe all the species according to each attribute, we will necessarily repeat the same things about many of them. (P.A. 639a 26-28).

The charge is that it would be repetitious. This amounts to a doctrine of simplicity as a criterion for theory evaluation. One can understand this in two ways: 1) psychological simplicity (which will cover linguistic simplicity), and 2) ontological simplicity. The first type would include reasons why presenting discrete descriptions and causes would be too difficult for the theorist. They would be unworkable _en masse_. The force of such an assertion is not that one way is _more correct per se_, but only that one method is practically workable, _while_ the other is not. If Aristotle is committed to realism (as G.A. 760b 29 suggests), then it would seem that workability is somewhat irrelevant to the truth of one over the other. Unless Aristotle states that it is absolutely impossible to proceed in one way (in which case one can never know the truth if the impossible way is also the true way), there is no force to the assertion that psychological considerations tell us anything about which theory is better.{3}

The ontological consideration states that such and such move based upon simplicity is justified because nature is really that way. If Aristotle can be interpreted in this way, then his view about proceeding from the common traits is equivalent to his stating that common traits actually exist in nature. If these common traits are ignored, then one is ignoring a fact of the natural order. It is not for mere psychological or heuristic purposes that common traits are examined. The biologist is to treat common traits because such traits are actually there. But _how_ is a common trait there? Aristotle describes it as a _general_ depiction, "for example, the locomotion of animals" (P.A. 639b 2). These descriptions are made from one description, the species, to another higher, more general description, viz., these common traits. After establishing that common traits can be laid down, Aristotle believes that the causes for these traits can be discovered (P.A. 645b) and classified (644b). The outlines to such a method are suggested in P.A. 1, 3 (to be discussed later). Basically all that is known at this point is that the biologist will take what is common to a group first, then consider the specific differences. These differences will be compared and contrasted so that links and ties between various species and genera may be established.{4}

The case for interpreting this passage as indicating ontological simplicity fits into other parts of Aristotle's theory. It also accords nicely with many modern notions of simplicity. But is this what the passage really says? Verbatim it seems as if Aristotle is advocating merely psychological or linguistic simplicity. If this is his position, it does not square with other statements (as already indicated). It also philosophically weakens the realistic force of any of his conclusions. However, to press the point of ontological simplicity, one must stretch the text and do a considerable amount of "reading in." I believe, along with Balme,{5} that Aristotle does not believe classification to be arbitrary (as psychological or linguistic simplicity could lead to), and that the various genera and species are not mere conventions but speak to the deep structure of the biological realm. This is the impression I receive from the balance of his biological writings. However, if that is what he means here it is odd that he did not say so. Could this be because, as Balme suggests, he did not yet have enough evidence?

The second question of major importance is:

> Should the student of Nature, just as the mathematician does when he is doing astronomy, first consider the datum of the animals and each of their parts and only then go on to state the reasoned fact and the causes, or should he follow another procedure? (P.A. 639b 8-11).

Should the student of Nature act as the mathematicians do when they do astronomy? How do these mathematicians act? First, it can be assumed that since Aristotle singles out mathematicians practicing astronomy one may rule out astrology or any other non-deductive method. Secondly, it is not only mathematicians, but mathematicians applying their calculations and procedures to physical phenomena. Now phainomena here seems not to refer merely to logical problems since he links phenomena to Nature and animals specifically, ton phusikon ta phainomena prōton ta peri ta zōa theoresanta (639b 9).{6} It is plain that Aristotle, here, is grappling with the types of questions outlined in the last chapter concerning the place of data in an explanation. Here we see a brief example of the two strains that Aristotle is trying to combine in his critical empiricism. This is a hot issue of the day as Aristotle reports to us that the answers to these questions "have not been decided" (639b 7). Whether he is right or wrong, Aristotle sees himself as a part of an ongoing debate. Some writers, such as the writer of

Ancient Medicine, are in the same camp as Aristotle (though clearly Aristotle's is the first systematic approach to the subject).

These two questions arise again in P.A. 1, 4 & 5. At 644a 25 Aristotle enlarges on the position he initially took. He declares immediately that it is species in which the theorist is interested. Presumably, the practicing biologist has to construct the various classes called species through some critically empirical process. But this point is not really discussed by Aristotle. He has said that one must begin with groups and proceed to the species just as definition operates through genus and differentiae. But it will be important to examine where these original groups are obtained. If Aristotle is to maintain his critical empiricism, these classes must be generated from individuals. The point may be put logically in modern terms by requiring Aristotle to employ both the criteria of class membership, as well as the criteria of class inclusion which his language usually suggests. Aristotle says:

> For since the "real thing" is what is indivisible in species, the most forceful, if it were possible, would be to make a separate study of each indivisible species, just as one studies "Man" or "Bird"; not just "Bird" as a group since "Bird" is a group which contains species, but the indivisible species of it, e.g., Ostrich, Cranes etc. But since this will result in speaking often about the same affection, because it belongs in common to more than one species, it seems long winded and unreasonable. (P.A. 644a 31-644b 5).

This passage seems to suggest that Aristotle thinks that studying each individual or even small groups is time consuming and causes undue repetition. {7} Repetition has already been shown to be irrelevant to a scientific method aimed at discovering the truth. The process is only unreasonable, if when proceeding thus, one grows no closer to the true picture of Nature, phusis.

The reason why this is confusing is that there seem to be two processes operating. In one process, the philosopher of biology has been given certain species, and he fits them into groups (genera). These genera, in turn, are placed into even larger groups (H.A. 490b 7). Once this has been discovered the philosopher of biology can start with the broadest groups and go back down observing the differentiae (P.A. 645a 35, cf. H.A. 486a 20, P.A. 644b 10, Metaph. 1018a 12, 1058a 7, 1037b). The second process involves getting the species to begin with. These groups Aristotle accepts from popular usage (P.A. 644b 3). As a philosopher of biology, Aristotle will accept the common names, such as 'fish' and 'birds' as representing genera. By doing this Aristotle must think that he is being reasonable in deferring to common opinion. If these various groups have been seen to adhere together, then perhaps there is a real class containing these members. By making such a choice, Aristotle loses the opportunity to establish a technical vocabulary which would prescribe what these groups would be called as well as offering more precise membership criteria.{8} With technical definitions (such as he created for logic), he could speak more exactly about species and genera.{9} This becomes especially important in difficult cases in either taxonomy or classification.{10} It should be noted that while Aristotle is restricted by this vocabulary, he is not blinded by it. He is aware of its restrictions as will be shown later.

In Aristotle's defense on this point one may probably point to his empiricism to explain why he accepted groups already popularly recognized. If he began with his own definitions, it might appear that Aristotle was imposing his own artificial groups upon the animal world. Being artificial, these groups would not be arising from the data, but rather would be an _a priori_ (uncritical rational) grouping. Such a system would be open to the kinds of searching questions Aristotle asks of the practitioners of dichotomy (P.A. 1,3). Besides, science must accept a certain amount of common opinion on various issues, since; if one were forced to establish everything from scratch, very little would ever be accomplished.{11} Natural group names were available to Aristotle, and he saw no reason to reject them, but assumed that we all can recognize certain natural groups which belong together. These requirements for class membership are from common experience. It is these first groups that are the indivisible species (_atomon eidē_). Here, there is a certain resemblance to the evolutionary biologists' term 'taxon' (pl. taxa).{12} Of all the groups, it is the most "real" because it is the only set whose members are individuals and not members of other sets.

The common opinion that establishes these first order groups is not the same as the common opinion, _endoxa_, mentioned in the last chapter.{13} In that case, the common opinions gave rise to intellectual problems which then became the object of inquiry. But Aristotle's point in _De partibus animalium_ is surely different. He is saying that natural groups do exist. Proof of this fact is that in ordinary language people make class distinctions through such terms as 'fish' and 'birds.' Presumably, this usage of language arose through someone noticing certain common traits and lumping many organisms into different "kinds" which were then given appellation. That these groups really represent organisms which belong together is apparent because the words are widely accepted. This illustrates wider empirical verification of those original usages.

But is this really a very good procedure? What if, for example, a dolphin were classified as a fish by many people because it lived in water and swam about. These "facts" might be observed by many and thus make their way into popular usage. It might be commonly recognized that a dolphin was a fish, but does this have any bearing on whether a dolphin should be represented as a fish? Certainly not, but perhaps Aristotle's point is something rather different. Just because he accepts the groups called 'Birds' and 'Fishes' he is not obliged also to accept all the members that may be ascribed to those classes, but only to accept that such classes exist.{14}

The original terms which were borrowed from popular usage can now be seen to be extremely vague. It is incumbent upon Aristotle to sharpen the definitions of these groups. In order to do this, he needs criteria for class membership. This explains Aristotle's emphasis upon examining common attributes of many species in order to establish appropriate "common" terms.

Once these general or common characters are determined, then other species are placed within the genus.

> Regarding the groups, which have been satisfactorily marked off by popular usage and possess both a single nature in common and species not far separated in them, such as "Birds" and "Fishes," and any other such group, even if they be without a name, yet contains species which are generically similar; it is correct to speak of some common affections, but when they are not like this, e.g., "Man," to speak of particulars. (P.A. 644b 1-7).

Thus, if some species were left out of a group, then inclusion in that group would depend upon matching the characteristics of one against the characteristics of the other. From this, one may also suppose that the biologist would be justified in excluding a species which varied greatly from the common nature of the group.

Aristotle is making physical qualities the basis for classifying animals (P.A. 644b 8). The first step in the process is establishing very general names to stand for these groups (via imperfect induction from particulars) as viewed through popular usage. Step two is to study these groups more systematically by finding the traits common to the various species such as Man, Ox and Lion. On the basis of step two, the biologist in step three can determine the common causes and then construct more rigorous criteria than were available in step one. Also, one will have an idea about various relationships that exist between groups and allow the construction of covering the species involved.

This is supported at P.A. 640a 12 when Aristotle presents two ways of practicing biology. In the one, he says a person can start with causes by describing the process of formation and then proceed to the results of the process (similar to the method of uncritical rationalism mentioned in the last chapter). The alternative is to start with the phenomena as they are and to describe the common attributes and then their causes. The first alternative is incorrect because the genesis of x cannot be known before x is known, since knowledge of x will condition the critical empirical explanation of x's genesis.{15}

Aristotle believed that this work had to begin with observations and so exhorted the potential biologist to get close to his data (something that astronomers, who study the heavens, cannot do).{16}

> Of all naturally composed things, some are ungenerated and imperishable for all time, and some are brought into being and perish. The former are of the most worth and are divine, but the opportunities for studying them are less, since there are not much sense data available for us to consider them. We have better information about things that perish, e.g., plants and animals, because they grow beside us. Much can be learned if one is willing to take the time . . . since it is possible for us to acquire more and better information about things in Nature (on earth), our knowledge of them has the advantage over the other . . . so it remains to speak of animals and their Nature. As much as possible we will not leave any of them out whether of lesser or greater value. For there are animals that are aesthetically unpleasing to the eye, but to the eye of science, those who can learn the causes and are naturally lovers of wisdom, Nature, which fashioned, them provides joys which cannot be measured. (P.A. 644b 25-30, 645a 2-3,

8–12).

Nothing can be overlooked, however mean and trivial it may appear to be. It is from this attention to detail that the whole emerges along with its causes (and purpose 645a 24, 35). Aristotle makes it clear that the philosopher of biology should demand a theory that is empirically based as he sets down a methodology which the practicing biologist may follow.{17}

2.3 GENOS AND EIDOS

Before proceeding very far into Aristotle's grouping of animals one must decide how the words _genos_ and _eidos_ are used by Aristotle. Now in a broad sense these words have been taken to be equivalent to 'genus' and 'species.' However the exact manner in which these terms should be understood has been a subject of much discussion over the past two decades. Important contributions on this debate have been made by A.C. Lloyd, Richard Rorty, Marjorie Grene, David Balme, Anthony Preus and Jim Lennox.{18} The force of these writers has been a further examination of just how these words are used by Aristotle. The problems are roughly twofold: first, does Aristotle mean something like class inclusion when he is using _eidos_ and _genos_, with the former being included in and not _indentical to the_ latter (in this case, the terms could retain their traditional translations of 'species' and 'genus'); or does Aristotle use these words in a loose way in which they are synonymous with a class of the same "type"? Second, when Aristotle speaks in the _Metaphysics_ of _genos_ as matter (_hulē_), what does he mean and how does this contribute to a quasi-phylogenetic relationship?

These questions will be addressed together. The first point to examine is the claim that Aristotle rarely (if ever) uses _eidos_ and _genos_ to refer to parts and wholes. Excluding _P.A. 1_, David Balme has noted 413 uses of '_genos_' and 96 of '_eidos_' in the biological writings. Of the 413 cases of _genos_. 275 are applied to a type of animal, 79 to animals in general and 25 to other kinds of things (e.g., _genos phuton_). The remaining 34 mean "kind" in an abstract sense including sometimes the sense of genus versus species. So in 354 cases, _genos_ refers to a kind of animal, a role which one might have expected to have been held by _eidos_. In 96 usages of _eidos_, only 24 refer to kinds of animals. Thus Balme concludes, "The traditional interpretation assumed that Aristotle did actually classify animals into genera and species, but this assumption is not supported by the evidence" (Balme, 1962, p. 85). Eidos, according to Balme, represents 'form' and has no separate usage to denote species, hence; no taxonomic distinction can be made between the two (ibid., p. 87).{19}

Lloyd and Rorty assert that the relationship between the two is as form is to matter. To do this they analyze Delta 28 (1024a 29-b9).

> The term 'genus' (or 'race') is used: a) When there is a continuous generation of the same type; e.g., 'as long as the human race exists' means 'as long as the generation of human beings is continuous.' b) Anything from which things derive their being as the prime mover of them into being. Thus some are called Hellenes by race, and others Ionians, because some have Hellen and other Ion as their first ancestor. (Races are called after the male ancestor rather than after the material. Some derive their race from the female as well; e.g., 'the descendants of Pyrrha.') c) In the sense that in formulae the first component, which is stated as a part of the essence, is the genus, and the qualities are said to be its differentiae. The term 'genus,' then, is used in all these senses— a) in respect of continuous generation of the same type; b) in respect of the first mover of the same type as the things which it moves; c) in the sense of material. For that to which the differentia or to which quality belongs in the substrate, which we call material.{20}

Meanings (a) and (b) can be discarded as irrelevant in this contest. That leaves the last two meanings which are summed up as "to d'os hulē." Other references describe genos as matter or link it to matter (cf. Metaph. 1058a 7, 1038a 6).{21} But as Marjorie Grene has pointed out there are other passages in the Metaphysics which seem to suggest a difference between genus and matter.{22}

> Things are said to be generically different whose proximate substratum cannot be analyzed into the other nor both into the same thing (e.g., eidos and matter into different genei), and things which belong to a different category of being. For on the one hand, some things signify the essence (ti esti) of being, while on the other hand, some signify a quality—others the various categories which already have been distinguished. These also cannot be resolved into one another or into some one thing. (Metaph. 1024b 9-16).

In this passage eidos and hule differ in genus so that hulē and genos do not appear to be synonymous. Other passages support this latter reading, such as 1015b 33 (where genus is 'in' the substance as opposed to the hexis of the substance) and 1071a 37 (where genos is discussed prior to the discussion of hule) and 1016b 33 where there is a clear distinction made:

> Again there are things which are one according to their number, others according to their eidos, others according to their genos, others analogically. Numerical refers to those whose matter is one; eidos refers to being one in definition (logos); genos refers to those who have the same figure of predication. "Analogically" refers to being related to another as to another.

It seems clear that at times Aristotle does refer to genos as matter while at other times they are distinct. This suggests that either Aristotle is confused or that he is up to different projects in the different contexts. I would like to suggest the latter possibility since it seems to me that Aristotle was indeed engaged in several enterprises at the same time: 1. He was making a technical distinction between two sets (genos and eidos), with one (eidos) being the proper subset of the other (this follows logically from his theory of definition); 2. He was toying with a concept of "quasi-phylogeny" and 3. He needed a base for #2 so he sketched out one possible explanation of it, viz., the genos-as-matter theory. These questions will be addressed, beginning with the first point while the second and third points will be dealt with together.

There are some obvious difficulties with the usage of genos and eidos, but most of these difficulties will disappear when one realizes that genos and eidos: 1. Have certain meanings when used singly; 2. When used together they logically relative to each other. This indicates that under the former usage, genos may be used to represent dog or any other "kind." Logically (under the second usage) this term moves upward and downward to fit terms more general or more specific. The cases cited by Balme in which genos represents a dog (658a 29) or again a kind of dog (H.A. 574a 16) are instances where genos is used alone to illustrate a particular class of animals. These cases are not used to differentiate a class and a subclass. For when used alone, genos tends to its popular usage of being a 'kind.' Distinct from its usage with eidos, this "kind" could be anything, nor is there any reason to think that this violates Aristotle's method. The only problem here might occur when one places the Linnean sense of genus upon Aristotle's genos. This would be incorrect for the Linnean genus has a perfectly stable meaning. Now it is used in a biological hierarchy of phylogeny. To expect this from Aristotle would be a mistake since he intended no such system. Eidos need not have a completely stable meaning as a level in a grand classification scheme (though the use is not completely random). The biological usage is derived from the logical usage in which the terms are related to each other as relative parts and wholes. In those logical cases all that is proposed is that within certain contexts, eidos, when used with genos will signify a part of a larger whole. One need not be surprised if eidos frequently has another usage when the term is used without genos. This usage is the common meaning of "form." As 'form' it can refer to the same entities that genos does in its common meaning of 'kind.' When genos and eidos are used together, then the former is a class and the latter is a subclass. Below eidos is atomon eidos. Thus Aristotle has three distinctions he can make. They will vary according to where one places the genus. For example, in P.A. 644a 35-644b 5, the genos is Bird; the eidos is unnamed; and the atoma eide are Ostrich, Crane and the like. Whereas the ultimate species (eschata eide) at 644a 24 are individuals, Socrates and Coriscus, within the species, Man. It is not of critical importance that genos may be used to describe an abstract group at times or a specific class, such as dogs, at other times. When used singly the terms take on their ordinary usage, but when used together these terms are logically relative to each other. The genos is a class and the eidos is a subclass (with the ultimate species being a subclass of the eidos). It matters not where one begins. Aristotle's logical classification can break up the problem into two or three parts.

Now Balme does point to one passage in which he believes that genos is used with eidos in a way inconsistent with the argument given above. H.A. 490b7 γένη δὲ μέγιστα τῶν ζῴων, εἰς ἃ διῄρηται τἆλλα (8) ζῶα, τάδ' ἐστίν, ἓν μὲν ὀρνίθων, ἓν δ' ἰχθύων, ἄλλο δὲ (9) κήτους. ταῦτα μὲν οὖν πάντα ἔναιμά ἐστιν. ἄλλο δὲ γένος (10) ἐστι τὸ τῶν ὀστρακοδέρμων, ὃ καλεῖται ὄστρεον· ἄλλο τὸ τῶν (11) μαλακοστράκων, ἀνώνυμον ἑνὶ ὀνόματι, οἷον κάραβοι καὶ (12) γένη τινὰ καρίνων καὶ ἀστακῶν· ἄλλο τὸ των μαλακίων. . . . . (16) . . . τῳν δὲ λοιπῶν ζῴων οὐκέτι τὰ γένη μεγάλα· οὐ γὰρ περιέχει (17) πολλὰ εἴδη ἓν εἶδος, ἀλλὰ τὸ μέν ἐστιν ἁπλοῦν αὐτο οὐκ (18) ἔχον διαφορὰν τὸ εἶδος, οἷον ἄνθροπος, τὰ δ' ἔχει μέν, (19) ἀλλ' ἀνώνυμα τὰ εἴδη . . . (31) τοῦ δὲ γένους τοῦ τῶν τετραπόδων ζῴων καὶ ζῳοτόκων δἴδη (32) μέν ἐστι πολλά, ἀνώνυμα δέ· ἀλλὰ καθ' ἕκαστον αὐτων

The main problems seem to occur in lines 16–19, "*periechei polla eide en eidos* . . . " Now it is important to remember that in this section Aristotle is trying to distinguish certain large groups of animals, such as Birds, Fish, Blooded and Non-Blooded. There are both "regular" genera, i.e., Fish, Birds etc., and "large or extensive" genera, i.e., Blooded and Non-Blooded. Within these classes are *eidē*. But as seen in the section discussed from P.A. 644a 35, *eide* can be comprised of sub-species, i.e., *atoma eidē* or *eschata eidē*. It is true that the two forms of *eidos* are not clearly distinguished here, but if *eidos* is taken as a logical subclass, then the expression in ll. 16–17 becomes "one subclass (species) does not comprise many subclasses (species)."{23}

The point here is that classification is very difficult and this is especially true of subclasses which oftentimes have no names (again, perhaps Aristotle yearns for a technical, zoological vocabulary). Balme is incorrect (p. 90) in associating the *en eidos* with the *genē megala* of line 16 and the *polla eidē* with the *genē karkinon* of line 12. Since there are three levels of discourse, it is just as fair to see the *genos* of line 16 as simply "other kinds not mentioned in lines 7–15." Thus, there are other genera, and classes and subclasses under them. Aristotle wants to say that these are difficult to name, and pose problems to anyone wanting to make a systematic classification scheme. The usage of genos in line 12 is relative to the large genera of line 7. In fact, with the addition of the *megista genē*, there are four levels of organization through two uses each of *genos* and *eidos* (i.e., 1. Large *genos* (Blooded and Non-Blooded); 2. *Genos* (Birds, Fishes); 3. *Eidos*; 4. *Atomon eidos*).

In other cases, there are types which do not fit into the scheme very well at all, like Man, who must be placed by himself. This same account explains the *eidē anomuma* of lines 31–32. This large group, viviparous quadrupeds, is comprised of groups and subgroups, but because Aristotle is relying on common parlance he does not have appellations enough for all the abstract levels. Aristotle's point is to emphasize the difficulties in classifying animals.

This is one part of the argument against the usage of *eidos* and *genos* as class and subclass ("genus" and "species" used relatively). There is still the question about whether *genos* is to be taken as matter and *eidos* as form only. The passages to support this thesis, it will be remembered, came from the *Metaphysics* (primarily Meta 1024a 29–b9). Of the four usages, the first two were found to be irrelevant. The third usage conforms to the part-whole relation described earlier. In no way can mathematical constructs be material in any but a metaphorical sense. I believe that much of this problem can be alleviated through emphasizing the logical force behind genus and species and differentia.{24} Genus and differentia are used in constructing definitions (Topics 101b 15 ff.). Surely this is the sense Aristotle is implying in the passage from the Metaphysics by using "logois" and "ti esti" (Metaph. 1024b 6). Aristotle is making a logical point which is not unrelated to his usage of genos as 'class' in the biological works. For when one gives the general properties of something, such that it is describing traits common to a number of things, then one is constructing a class in which differentiae may "pick out" the item in question. One can trace Aristotle's biological use to genos and eidos to his logical use of genos. It is this use of genos that is being referred to in line 5. However, there is still that troublesome statement in line b9 that 'genus' has "The sense of material. For that to which the differentia or quality belongs in the substrate, which we call material." But again this must be meant analogically since Aristotle explains this sentence three lines later with the illustration of form and matter differing in genus. If matter differs from form in genus, then surely genos and hulē are not the same.

Two other texts must also be examined to secure this reading. The first occurs at 1038a 5-8, "If then the genus absolutely does not exist apart from the species, which it included, or if it exists, but only as matter (for speech is genus and matter, and the differentiae make the species, i.e., the letters out of it), obviously the definition is the formula composed of the differentiae." What does it mean to say that the genus exists "just as matter" (hos hulē)? This same usage is displayed at 1058a 21-24.

> It is clear that to what is called genos no eidos is identical to it nor other in eidos. This is proper, for matter is revealed through negation, and genos is the matter of that which is called genos, not as the clan of Heraclidae, but as we speak of genos in nature.

Reading legetai as "is called," there appear to be two distinct usages of genos. "Genus is the matter of that which is called genus" means genus is the matter of that x (called genus) as an element in its nature. But matter is disclosed only by negation, i.e., (reading with Ross){25} by eliminating the form in the concrete thing. Therefore, genos also is disclosed by negation, eliminating the form in the concrete thing and taking it to a higher state of generality. Having no form, it exists only potentially as a logical component in any actual genus in nature. This potentiality is undifferentiated matter and does not exist (Metaph. 1029a, 1044a 15, 1045a 5, Generation and Corruption 319a 26, P.A. 643a 25). It is a component to any genus in nature. This potentially is also logically a genus. Just as matter is a precondition for any particular thing (as is form), so each of them (form and matter) constitutes logical categories of supreme generality, and as such, are genera for defining what comes next. For example if one were to talk about the contraries hot, cold, wet and dry, he would need to appeal to broader categories which could be said logically to generate them. These logical categories would be the genera: form and matter. They are genera in a logical sense (as well as a material sense which we will see later). They act as a precondition for any existent genus.

The confusion in this passage comes from the two uses of genos. The first refers to a logical class, while the latter refers to a kind in nature. Just as eidos has different uses for Aristotle, genos does also. That Aristotle has two senses in mind is shown in his example of 'clan' (of Heraclidai)--the class--and 'kind' in nature. His first use of genos is as class, while the latter is a kind in nature.

Aristotle had logical categories in mind when constructing genos and eidos as indicating a part-whole relation. These relations that are developed in the non-biological words carry over to the biological organization as follows.

1. The individual--one in number; matter identical with itself.

2. Different individuals in one species--vary in number; form the same.

3. Different species in one genus--vary in matter by "more and less"; function is the same.

4. Incomparables--likeness only by analogy.

Now there is a way in which this chart seems to pin down the usage of these terms. As a systematic biological classification these added restrictions appear to point to definite levels of organization. That is, genos is a group in which the parts perform the same function. When these parts differ in kind, but perform the same function, then they are generically different. They are similar only with respect to some common function. Thus the function of man's lungs is to breathe, and the function of a fish's gills is to breathe. They share a common function, and so in that respect are similar. But the actual parts which perform these functions are not similar in construction or shape. This is what it means to differ by analogy.{27} The definition is clear. The nails of a cat and the nails of a lion cannot be said to differ by analogy so that one could not, by this standard, make a cat and a lion separate genera. Parts which differ by function and shape and structure are all together different. Most parts of a man and a sponge differ in this way.

Parts which differ only slightly and perform the same function differ by "more and less." This constitutes a difference in degree. An example of this would be the wings of a hawk and the wings of a dove. These differences characterize animals differing in species.{28} All of this seems to tie down the uses of <u>genos</u> and <u>eidos</u> in ways not observed in the biological works themselves. To an extent this is true. Aristotle, in his philosophy of biology, was laying the theoretical foundations of taxonomy.

It is quite natural that Aristotle might apply his terms in the biological works themselves differently than he did in the theoretical works. He was addressing a different task. Just because he may have laid a foundation for how classification might be done does not mean that he intended actually to carry it out himself. His different uses of <u>genos</u> and <u>eidos</u> may be accounted for because they allowed <u>genos</u> and <u>eidos</u> in one context to have an exact classificatory meaning. Perhaps Aristotle is showing his awareness that any system of classification must have stable meanings and not merely have relative levels of organization. Whether or not he thought this, there is still a sense in which even this apparent attempt at specificity contains some of the relativity of his logical use of these terms.

This can be seen through the practical problems with applying his criteria to determining whether two groups vary as species, genera or all together. For example, if one compares the legs and feet of a cow and a lion, it is apparent that the legs differ by "more and less," while the feet differ by analogy. The stomachs also differ by analogy, while the lungs differ by "more and less." How does one ultimately determine whether two unknown animals differ specifically or generically?

Does one make a judgment on each part and function and then add up the balance sheet?

Aristotle is not specific on this point, nor does one expect that the problem bothered him. His aim was not to create a worable plan in its final form. These biological categories are tentative proposals derived from his logical usage of _genos_ and _eidos_. As logical categories they are basically relative in their usage. As biological categories Aristotle is suggesting a system of class inclusion which might be constructed for organizing animals. In a third context, that of his actual biological investigations (as well as other non-technical usages in his works at large), Aristotle reverts to the common usage of _genos_ as 'kind' and _eidos_ as 'form.'

Having established that _genos_ and _eidos_ do refer logically to sets and subsets, let us examine Aristotle's usage of these terms as they apply to Nature, viz., _genos_ as _hule_. Jim Lennox has written persuasively on this point arguing that,

> . . . the genus-term in a definition picks out a _shared substratum_ that is differentiated differently from one species to the next. I recommend viewing the genus as a unified system of shared characteristics (e.g., winged, two-legged, beaked, feathered, oviparous), all of which are possessed by each of the members of at least two species.{29}

Lennox states further that genus [system of shared characteristics] is the "material" for specific differentiae in three ways:

> 1) each specific differentia is an (but not _the_) actualization of a potential possessed by the generic character; 2) every generic character, and each specific character implies the existence of the generic character; and 3) our ability to identify genera indicates that the generic characteristics are still apparent in their differentiated state" (p. 337).

Lennox's reading, it seems to me, correctly points out one feature of Aristotle's enterprise. Aristotle was trying to give a physical explanation for why certain animals appeared to be in "natural" groups. What is 'natural' about these groups? Is it that they are only picked out randomly or according to the perspective of the observer? Is there something _objective_ about a method of grouping by shared characteristics? The answer is yes, and this objectivity arises from discontinuous genera which are differentiated (at least in theory) on a specific continuum. This genus thus acts as a common substratum for its species (i.e., as substratum, hupokeimenon, it functions as matter). This is the second aspect of genos. It is the sense which renders this brand of systematics as explanatory since the explanans clearly offers an ordering principle to the explanandum (though it is important to realize that the second usage of a kind in nature "as matter" depends upon the logical meaning of the first usage, viz., that of set and subset, in establishing this latter point).{30}

It is in this way that Aristotle can show why x and y belong to one genus and not another. They both possess shared character states. These character states are of an entity, the genus, which thus has an existence-- it exists through the species. The indivisible species manifests both the genus and the species.

From this, two points become clear: 1. The definition of the indivisible species is not arbitrary for it picks out actual traits which are derived from more general classes{31}; and 2. The genus is no mere abstraction, but exists actually through the existence of members of some indivisible species.

Some parallels might be made here to modern phylogenetic systematics or 'cladistics.'{32} It is true that Aristotle's use of eidos and genos cannot be said to infer phylogenetic groups, as 'phylogenetic' implies biological evolution which Aristotle did not endorse. But, one could speak of quasi-phylogenetic groups since Aristotle does suggest a theory of synapomorphy. 'Synapomorphy' is the sharing of a derived character state or a later stage in the transformation sequence of a derived character state. This provides explanatory grounds for the construction of "cladograms." In an analogous way, Aristotle's use of shared-derived character states is explanatory (in a non-evolutionary manner) of certain relationships which exist between groups of animals.

2.4 DOES ARISTOTLE HAVE A CLASSIFICATION SYSTEM FOR ANIMALS?

Now that an adequate understanding of genos and eidos has been established, the question remains how far Aristotle went in applying this to construct a system of classifying animals. From the last section it is clear that Aristotle had some notion of biological ordering, but just how far he meant to take it is another question. There have been some writers, such as Charles Singer, who have attempted to reconstruct tight zoological hierarchies from Aristotle.{33} Singer goes further, constructing a <u>scala naturae</u> on the basis of two passages, H.A. 588b 4ff. and <u>P.A. 681a 15ff</u>. In these passages Aristotle states that nature, "Proceeds little by little from inanimate things to animals in such a way that it is impossible to ascertain the exact point separating the two, nor on which place an indeterminate form should lie." The passage from <u>De partibus animalium</u> indicates the minute gradiations between plants. Others, like Norman Platnick, have constructed cladograms to illustrate the relationship between various genera and species.{34}

These are important passages, but do they indicate that Aristotle had a comprehensive classificatory scheme in mind? On the contrary, they point to the difficulty in establishing such a scheme. Add to this the fact that Aristotle's use of <u>genos</u> and <u>eidos</u> are only relative, and the picture becomes more difficult. Some attempts have been made to sort out these various generic systematizations by Jurgen Meyer, and Aubert and Wimmer,{35} but even their exhaustive work cannot fit together the many disparate 'kinds' mentioned in Aristotle's biology. D.M. Balme has shown that of 550 animals cited by Aristotle, there are 60-65 groups having about 200 members. This leaves over half of the animals unaccounted for. Even worse is the fact that of these 60-65 groups, many are subdivisions of common types, such as different varieties of dog or hawk. The real problem is to be able to find intermediaries between the large genera and the more common groups. Meyer offers the following list.

1. <u>Tetrapoda zōotoka</u>: <u>lophoura</u> (horse, ass, mule); <u>pithēkoeidē</u> (apes and monkeys).

2. <u>Tetrapoda ōtoka</u>: [lizards]

3. <u>Ornithes</u>: <u>gampsōnucha</u> (birds of prey); <u>steganpoda</u> (swimmers); <u>machroskelē</u> (waders); <u>epigeia kai ptētika</u> (gallinaceous).

4. <u>Kētē</u>: [wales, dolphins]

5. <u>Ichthues</u>: A. <u>selachē</u>--(i)<u>galeōdē</u> (sharks); (ii) <u>kerkophora</u> (rays); (iii) <u>batrachos</u> (fishing frog); B. <u>ōtoka</u> (all other fishes)

6. Malakia: [cephalopods]

7. Malakostraka: [crustacea]

8. ostrakoderma: strombōdē (snails); dithura (mussels and oysters); monothura (limpets).

9. Entoma: koleoptera (chafers); melittōdē (bees and wasps); diptera (flies and gnats); makra kai polupoda (centipedes).{36}

Now Aristotle is not consistent in following his own categories and rules for classification so that it seems that his descriptions are more contextually motivated than prompted by some desire to create a grand classification scheme.{37} One can conclude from this that Aristotle was a total failure and incompetent (if as Meyer says, the list of 550 animals is only legitimately broken into 15 groups, comprising some 100 animals) or assume that Aristotle was up to something else. Now is it true that Aristotle nowhere claims that he has a system by which he can classify all animals. But he does offer certain groups and the outlines of a broad scheme. What then is his position? Does Aristotle want to classify animals, and if so, then why are the outlines of his system so sketchy? Perhaps one can account for this by pointing to the limited observations of the Greeks compared to peoples less civilized than themselves. As Charles Singer says:

> It was the world in relation to himself and not as a mere objective complex of phenomena that interested and appealed to the Greek. Thus in his attention to purely external phenomena he ranks no higher than many peoples infinitely his inferiors in other respects. It is said that any Malay can distinguish some three hundred species of insects; set off against this the whole fauna of Aristotle with its total of but five hundred and forty forms. Again, Theophrastus, a professed botanist, mentions only three hundred plants, practically all cultivated; compare this to the record of barbarian Anglo-Saxon speaking tribes of whose literature the merest fragment survives, but a fragment containing about eight hundred plant names.{38}

In the end, the question is actually resolved by returning to a point made in the first section of this chapter. Aristotle in P.A. 1 and H.A. 1,1-6 (where these discussions about classification take place) is engaged in philosophy of biology and not in practicing biology itself. As such, he is perfectly within his province in offering an outline of how one could construct a system of classifying animals. The question thus becomes, "How ought one to proceed in classifying animals?" and not "Where is Aristotle's finished system"? To say that Aristotle never produced a finished and complete system of classification is only to say that qua zoologist Aristotle was either not up to the task or that as a practical endeavor the task did not interest him.{39} What is clear, however, is that Aristotle came up with a method for constructing a system of classification which was different from the non-unified system that preceded him.{40}

The exact nature of Aristotle's contribution may be sharpened if brought into the context of modern zoology. A survey of a few important terms will facilitate this.

Taxon, Taxonomy: A taxon is the concrete object by which the zoologist may classify. Such entities as birds, robins, and veterbrates are all taxa. As Simpson says, "A taxon is a group of real organisms recognized as a formal unit at any level of hierarchic classification."{41} Taxonomy refers to "the theory and practice of classifying organisms."{42} Some have wanted to make a distinction between the theoretical aspects of taxonomy and its practice,{43} but following Simpson and Mayr, I will use the single term to refer to both activities, with this distinction, that taxonomy may be further termed alpha and beta depending on whether the zoologist is describing new species (defining new taxa, alpha) or whether he wishes to reorder the existing groups, beta (Simpson, 1961). This traditional line puts beta taxonomy very close to classification. For the purposes of this essay, beta taxonomy will thus refer to the theoretical considerations involved in altering a method of classification (i.e., placing of taxa into classes and the creation of those classes).

Identification: Identification is undertaking to place individuals into taxa according to a deductive procedure. In this process one begins with a given set of taxa and tries to fit the specimen before him into one of them. If one is successful, then he has identified the specimen (cf. Mayr, 1969, p.66).

Classification: This domain covers the grouping of taxa within a set. The real objects are the taxa and the set in which they are placed is the class. This, of course, may be relative according to the level on which one is operating. For example, wolves, foxes etc., may be considered as taxa and combined into the class (family) Canidae. In this operation Canidae is considered as the class. But when grouping Canidae with cats (Felidae), bears (Ursidae) and other predacious mammals into another class (the order Carnivora), then Canidae is considered as a taxon. In short, the members forming the class are the taxa, while the class is that in which they are placed. Classification is thus an inductive grouping of taxa on a particular level of generality.

Systematics: This can broadly be defined as "the scientific study of the kinds and diversity of organisms and of any and all relationships among them" (Simpson, 1961). More precisely, systematics will refer to the broad, general concerns that the theoretical zoologist or philosopher of biology may have in ordering the organisms of the world. This involves certain fundamental approaches as to how and by what principles a system of classification should be constructed, and *a fortiori* how beta taxonomy (the logical extension of this activity) ought to proceed. In this way systematics can be viewed as "meta-beta-taxonomy."

With these terms well in hand we can proceed to a brief overview of the three major schools of systematics and their resulting systems of classification. The first and most entrenched school is evolutionary systematics. This school claims that life has evolved and the goal of systematics is to reveal that history as accurately as possible. Since evolution implies ancestry and descent, the system of classification should reflect this as much as possible. The mode of beta taxonomy is called "upward" since one moves from the specimen to the appropriate species and sub-species and then to the more general classes on the phylogenetic tree. This is done by resemblance which may be termed: a) true evolutionary resemblance acquired by a common ancestor and b) false resemblance (often called convergence—"parallelism" is midway between (a) and (b)).{44}

The second variety of systematics is numerical taxonomy or phenetics. This model was pioneered by Sokal and Sneath.{45} These writers' central methodological principle is phenetics which occurs from the clustering of samples ("operational taxonomic units" or "OTU's"). These are computerized indices of the overall similarity of biological forms according to an algorithm specificing various parameters. The data of these are used to construct phenograms. All this is done in the spirit of methodological objectivity. In numerical taxonomy no distinction is made between "true" (phylogenetic) and "false" (convergent) resemblances. Instead of constructing phylogenetic trees, the numerical taxonomists are content with statements of the relative nearness of samples.

The third school is called phylogenetic systematics or cladistics.{46} This school seems to be between these extremes. Like evolutionary systematics the cladists accept the distinction between true and false resemblances. They also believe in constructing phylogenies. But then they further add to the normal distinctions that,

> For any monophyletic taxon (i.e., a taxon composed of two or more species consisting of an ancestral species and all its known descendants), evolutionary similarities shared by a group are of two sorts: Those held over from some remote ancestor (e.g., two pairs of limbs in mammals vs. those held only by members of that group (e.g., three inner ear bones in mammals).{47}

These two traits are termed synapomorphy and autapomorphy respectively and multiply the number of similarities recognized by evolutionary systematics. As with phenetics the cladists stress phenotypic traits and do not emphasize the genotype. The major principle of their systematics is descent due to chronology. Thus, for example, taxa in a cladogram (the diagram which marks off such divisions "are subordinated to one another according to the temporal distance between their origins and the present; the sequence of subordination corresponds to the 'recency of common ancestry' [of the taxa]" (Hennig, 1966, p. 83). On these principles, taxa which originate from a split in the Precambrian period are ranked as phyla; between Cambrian and Devonian as classes; between Mississippian and Permian as orders; between Triassic and Lower Cretaceous as families; between upper Cretaceous and Oligocene as tribes and in the Miocene as genera (cf. Hennig's figure 58 on page 186).{48}

One can see how this emphasizes geneology which is determined by phenotypic traits. The evolutionary biologist, on the other hand, would claim that genetic considerations must play the foremost role. This causes a conflict between the two.

> The argument of the cladist fails to recognize that the term relationship has two distinct meanings, genetic relationship and genealogical relationship. The two happen to coincide for all practical purposes as long as we deal with close relatives . . . . In phylogeny, where thousands

> and millions of generations are involved, thousands and millions of occasions for changes in gene frequencies owing to mutation, recombination, and selection, it is no longer legitimate to express relationship in terms of genealogy. The amount of genetic similarity now becomes the dominant consideration for a biologist . . . if one of the lines is exposed to severe selection pressures and as a result diverges dramatically from its genealogically nearest relatives, it may become genetically so different that it would be a biological absurdity to continue calling them near relatives. Even though the crocodilians are cladistically nearest to the birds (both having descended from the pseudosuchians), the crocodilians are still closer to many of the other reptiles, as far as the total gene composition is concerned than they are to the birds, which have so drastically altered their genetic composition.{49}

With this brief background into modern systematics, let us return to Aristotle. What is *he* up to? To begin, as has been already stated, Aristotle does *not* hold an evolutionary theory as Empedocles did. Aristotle saw Empedoclean mechanical explanation as fraught with difficulties (see Chapter Five), and did not see any reason to posit anything other than that species are continually existing entities (after all there were no *phainomena* to the contrary). This means that one of the primary explanatory tools for modern systematics (at least for evolutionary systematics and phylogenetic systematics) is absent in Aristotle. What then does Aristotle use to replace this void? The answer can only be stated in terms of Aristotle's actual program. Now, it is true that Aristotle does engage in some identification and alpha taxonomy particularly in the *Historia animalium* (see Chapter Four for some examples of this), but it is also true that this is not Aristotle's principal concern. Aristotle seems most interested in systematics. He wants to lay down principles for how one would go about organizing the diverse organic world. His method is to begin with natural groups. These groups have various sub-groups which are, in turn, composed by individuals (P.A. 644a 35). Through imperfect induction these groups have been commonly accepted. Now there are many other ways that animals have also been viewed, e.g., blooded, viviparous, wild, tame etc. One can make a large list, but if one only considers common attributes (644a 26), then certain larger groups emerge. Aristotle asks the critically empirical question, "What attribute is present?" Some traits serve to unite the "natural" groups into larger groups. These larger groups are generalizations of essential common traits (the quasi-phylogenetic traits of 2.3). That they be essential is important, since otherwise one would obtain a randomly chosen group. The test for these essential traits begins with sensory experience (645b) and is processed by reason: one first finds attributes, then one finds causes. The result is a hierarchy of classes with some common essential attribute. This is similar to the method employed by Linnaeus in his beta taxonomy.{50} Thus in this respect Aristotle was a philosopher of biology engaged in systematics. That his "system" seems incomplete as was stated above can be explained by the fact that he had no system. System building is the work of beta taxonomy. Linnaeus was interested in this endeavor, but Aristotle was not. In systematics one engages in "meta-beta-taxonomy." The philosopher/theoretical zoologist's interest is with how one would go about constructing a full-blown system of classification--not the actual building of one. This is the work of beta taxonomy. To criticize Aristotle for not doing a good job as a zoologist engaged in beta taxonomy is to misunderstand what Aristotle was doing.

One can see some similarities between Aristotle and all three branches of modern systematics.{51} First, and most obvious, would be the affinity with numerical taxonomy. Aristotle did, to an extent, put organisms and species together according to their morphology (see the next section for more detail on this). In a way he was not too far removed from phenetics for he puzzled over how this should be weighted in an overall assessment of the biological forms in question. But like the cladists he was interested in using these shared derived traits to create "quasi-phylogenetic" groups. They are "quasi" because though there is no true historical descent, there is a sense of synapomorphy and autapomorphy that can create groups which have at least logical descent through the application of "the more" and "the less."{52} This descent can (with reference to Aristotle) be termed "logical" since no ancestry is implied. But this does not mean that there are no physical manifestations (cf. the discussion of ontogenesis in G.A. 736b 2-5, where the embryo is said to pass through stages in which it is first an animal before it is a human fetus). On the contrary, there are physical effects among which are the generic characteristics that seem to affect the organism in ontogenesis.{53}

This would seem to provide a tie to evolutionary systematics, for this ontogenesis comes about from the union of sperma and katamēnia (G.A. 723a 20). As was shown in the last section the genos acts in this respect, just as matter, and therefore, provides a further explanatory tie to the existence of these characteristics which the zoologist observes.

However, too much can be made of these similarities. Though there are certain likenesses between Aristotle's systematics and modern theories, it must be noted that (except for numerical taxonomy) Aristotle's proximate causes for classification are considered only remote causes today. However, it is perhaps to be expected that Aristotle would have chosen attributes and their causes (as well as certain functions or activities, 645b) as a basis for his systematics through which he believed a classification could be constructed. After all, this is basically a critically empirical justification, resting upon observation and offering a coherent way that the natural diversity could be organized.

2.5 ARISTOTLE AND DICHOTOMY

Aristotle attacks dichotomy as a method of definition or classification because it is essentially an uncritical rational procedure which does not regard the "natural" groups as viewed empirically. Against dichotomy Aristotle offers six arguments. Some of these overlap, but it is useful to discuss each.

The first objection occurs at 642b 8. Division is unnecessarily repetitious. To define some animal first as having feet; then as an animal having feet but with two of them; then as an animal having feet, with two feet, and with cloven feet, is cumbersome.{54} It is only necessary to state the last differentia. The others are irrelevant. Aristotle repeats the same point in Meta. 1038a 19, "*he teleutaia diaphora he ousia tou pragmatos esti*." Unnecessary repetition violates the dictum of simplicity which Aristotle enunciated in P.A. 639a 26 (discussed in 2.1).

The force of this objection lies in the words, "*tauton pollakis anakkaion legein*." If repetition is unnecessary, then it is analytic that it be eliminated. However, what is to determine "unnecessary?" If repetition were necessary to arrive at an end unavailable by any other means, then repetition would seem to comprise no harm. However, it is not the repetition alone which makes dichotomy unwieldly under this objection, but repetition in view of the fact that only the final differentia is necessary to denote the essence of the thing in question. If the final differentia were all that was necessary, one might suppose that it should be stated from the outset without such a long process. But what if the final differentia could not be procured unless a process of division preceded it? If this were so, the practitioner of division could admit that the final differentia is the most important, but retort that this differentia cannot be obtained except as the result of a long division. By itself, Aristotle's first objection does not appear very strong and works only when the final differentia is obvious.

Aristotle's second objection fares better. Dichotomy is said to break up natural groups such as Birds (642b 10). Aristotle claims this was the practice of certain writers of his day who split up Birds by putting some with water creatures (presumably others were grouped with land and/or air creatures).{55} By such a division one could be grouping Birds and Fishes as water creatures, and Birds and Insects as air creatures. If there are "natural" groups, then this would be a mistake because these natural groups would no longer be together.

This can be a very strong objection to dichotomy if unpacked some. The point is that there are "natural groups." Where do these groups come from? How are they recognized? Why do x, y, z fit into one natural group and a, b, c into another? These questions might have seemed trivial to Aristotle. He operates under the assumption that we all know what a natural group is. It is a group like Birds. Is a bat a bird? If not, why not? How does one fit an unknown specimen into the system?

This, of course, is a problem in identification.{56} Identification is a deductive procedure. One assumes that the species and sub-species represent a fixed ordering of criteria which act as a "covering law" for that particular instance. The individual is subsumed under the appropriate category and is identified. This is done by comparing the specimen as observed under the dissection microscope to the "covering law" (termed by Simpson, 1961, pp. 183-186, as the result of a _hypodigm_ which constitutes the basis for the description of the species. It is this _description_ which is used in identification as the "covering law." The _hypodigm_ is the entire sample of specimens personally known to a given taxonomist—engaged in alpha taxonomy—at a given time and is considered to consist of unequivocal members of the taxon.) Now it is generally the case that no actual specimen is _exactly_ like the model. That is to say that there are phenotypic variations among individuals within a given specific population. Since this is true, what then do we make of these "covering law" models/descriptions of a species? Put more pointedly, the species model which one works with in identification seems to suggest a static type. The notion of a static "type" reflects the evolutionary systematics school's eschewal of an earlier notion that taxa (usually species) do not vary. This view is often attributed to Linnaeus,{57} and sometimes is incorrectly said to have originated in Aristotle.{58} The hallmark of this approach is that: 1. The essence of a species precedes the individual itself, and 2. Because of this, the names of these species (as they represent these essences) become of greater importance than the specimens (zoological objects). Thus there is created a fixed unvarying species (or other taxon) in which the practicing zoologist identifies specimens as "better" or "worse," relative to that fixed ideal. Of course, diversity within the taxon is recognized, but this diversity is usually viewed as a defect of the specimen according to how these defects alter the specimen from its type.

Aristotle is not a typologist in this sense at all. First, it has been argued that Aristotle was interested first in actual specimens and engaged in alpha taxonomy (for examples of his research see Chapter Four). This required him to engage first in imperfect induction before proceeding to perfect induction. The only cases in which Aristotle did not act in this manner were when he had an "accepted body of data" (which is a principle presently accepted as well; see the International Code of Zoological Nomenclature (1961) Arts. 65a, 70). This provided him with the imperfect induction groups. Secondly Aristotle did not put great emphasis upon names (as has been shown earlier). In fact, he often could have used a technical vocabulary. This was not Aristotle's concern. Finally, though Aristotle did believe species to be fixed and eternal, specimens were not said to be "defective" unless they could not perform the activities normally (statistically) ascribed to some group. For example, if a robin cannot fly and it is known that robins as a group are fliers, then this robin, alpha, may be said to be defective in just that way. That alpha cannot perform a "normal" (statistically verifiable) function means that it is defective with respect to flying.{59}

It is true that Aristotle did not hold specific diversity as a value, but neither did he denigrate it either. Aristotle was always the first to point out that if some specimen varied too much from known examples that it deserved its own group.{60} Diversity was acknowledged--not as a defect nor as an asset, but merely as a fact of Nature (this is assured by Aristotle's theory of inheritance see: G.A. 766b 15 ff.).

This last view seems to be consistent with Aristotle when he both admits that individuals are the most real (644a 25) and when he introduces his criteria for biological nomenclature, viz., that species vary by 'more and less' and genera by analogy. Aristotle does not contend that all individuals are alike, but he does contend that natural groups may be formed, and that within these groups there are fixed essences, e.g., the essence of Man is to be rational (P.A. 641b 7). Aristotle used the eidos of some system as its organizing principle. In this way he is ascribing essences to that system, viz., those qualities which make the system a system. In this respect Aristotle's biology is essentialist. The eidos is not a rigid unvarying type but a "normative type" with certain degrees of fluxation within it. However, it is essentialist and therefore antithetical to evolution. Several articles written recently have tried to minimize Aristotle's essentialism, e.g., Balme (1980) and Lennox (1981). What they really attack is the notion of rigid essentialism (which I also attack) being attributed to Aristotle. However, as Hull (1981) has pointed out, it is dubious whether any serious biologist in history would ever consider such a variety of essentialism. A more interesting premise to examine is whether, in fact, Aristotle was an essentialist at all. Lennox (1980), p. 345, illustrates a view which many contemporary Aristotelian scholars hold on the subject, "Aristotle does hold a brand of essentialism . . . there are certain functions, and consequent details of organization that are necessary for the 'life-style' of each species of a genus that determine the exact conformation of each generic character within that species. But this essentialism is not 'Platonic' or inimical to evolutionary biology. On the contrary, with its stress on the way in which requirements of adaptation determine the precise difference between species within a genus, it is a spiritual ancestor of the evolutionary approach" (cf. Balme (1972), p. 97, "There is nothing in Aristotle's theory to prevent an evolution of species"). It is true that there are strains of similarity in meta-beta taxonomical theory between Aristotle and modern evolutionary systematics. But surely this cannot be pressed very far. If Aristotle resembles any modern school of classification it would be either the numerical taxonomists or the cladists. These similarities, as I have shown, are much more apparent. To adopt the other view too strongly would require, as Hull says, that "Aristotle held none of the views commonly attributed to him during the past 2,000 years" (Hull, 1981, p. 8).

THE METHOD

Essentialism (in some form) has been a part of biological theory for some time. It certainly did not start with Aristotle, and some of the ancient commentators on Aristotle may have exaggerated it in their own writings. But there is an essentialism there, and to pretend it is not present is to be guilty of vicious revisionism.

It is my contention that the best way to understand Aristotle's essentialism is to examine eidos. Eidos, in turn, should be examined in terms of a biological system. A biological system is comprised of form and matter. Just what will count for 'form' and 'matter' is relative to the system in question. If the system is a codon of DNA, the form is the particular coding configuration of nucleotides, while the matter is the bases themselves (the purines and pyrimidines). If the system is a Man, then the form will be his total shape plus his nutritive, sensitive and rational souls. The matter in this case would be the various uniform and non-uniform parts.

These parts only become biologically interesting when they are organized (P.A. 645a 34). This organization of the system is its form, and this form is prior to its matter. As Marjorie Grene writes, ". . . Form is then prior to matter, not only cognitively but ontologically as well."{61} It is the form of a system which is prior to its matter because the system is only identified as organized, i.e., through its form or organization. Form is logically prior because matter would not be what it is without that particular form--being logically prior (P.A. 639b 25). The higher level controls the lower. Thus Aristotle says that a man has hands because he is rational and not the other way around as Anaxagoras claimed (P.A. 687a 5){62} Rationality as a part of the form of Man makes the matter what it is. This is analogous to Cuvier's claim that the genus bestows traits upon the species.{63} There is a sense in which part of the organizing principle of some system might be that it exists within a natural group. These natural groups thus become "formally" related because their systems are alike in their functions by aiming at similar things in similar ways, i.e., they are related causally because their various causes roughly correspond (cf. to the cladists here). This does not eliminate variation within a species or genus, but it does allow for "traits" to emerge. These characteristics, the form, can be truly applied to the genus or species in question so that if someone were to ask, "Why does that animal have feathers and wings," the appropriate reply would be, "because it is a bird."{64} And one answers likewise to a question concerning traits on the species level, "Why does that animal have black feathers with a red patch on the black part of its wings?" The response is made on the species level, "Because it is a red-winged blackbird." For general traits the appropriate tie is to the general category. As one enumerates traits more precisely, one's response also becomes more exact. In each case the characteristics in question fit in at some level of description. This description is true of that level as a set of traits which all things at lower levels do in fact possess. The same procedure is necessary in modern evolutionary taxonomy. Whether one talks of types or statistical populations, the practical result is the same. A systemic hierarchy is constructed based upon individuals' common traits which form abstractions of increasing generality. Though the notion of "gene pools" seems far removed from Aristotle, each theory operates under methods that are not all that different. Aristotle's categories are no less empirical than those of modern taxonomists. This is particularly true of the numerical taxonomist and cladist. But even for evolutionary systematics it is clear that Aristotle is not the _a priorist_ that some taxonomists have charged him with being, e.g., tagging species by means of one single character selected _a priori_ and abstractly.{65} Aristotle is striking out against just such systems of classification in his attacks on the dichotomists.{66} Though Aristotle's classificatory techniques are less sophisticated than those of modern taxonomists, Aristotle's method of considering many differentiae which might have a bearing on the total _eidos_ of the animal puts him near the moderns with respect to empirical techniques.{67} It is the very invocation of _eidos_ that actually emphasizes Aristotle's empiricism. For the determination of the _eidos_ (as "form") allows the creation of the taxonomical _eidos_ (as "species").{68}

Aristotle did not carry out his plan very far, as has been shown in the last section, but his principles for taxonomy were not radically altered until the inclusion of evolutionary theory into taxonomy (and even then, as has just been shown, there are a few striking similarities). Even at that, Aristotle's insistence on first studying biological forms (systems as organized) is still an important principle. Another key tenet is the priority of higher levels of organization over lower ones (though this is not accepted for evolutionary reasons). Both points combine to make Aristotle's contribution in this area significant.

A variation of the theory of 'types' is a theory of favored traits (cf. cladists). It must be agreed that to an extent some traits _are_ favored over others in systematics (e.g., 'true' vs.

'false').{69} The problem occurs when one has to decide which traits are to be favored. It would seem that there are no empirical reasons for making this choice beyond the rationale of natural groups via synapomorphy and autapomorphy—and these yield similar answers to both moderns and to Aristotle. Since it is impractical to treat every trait as equal (for then the sample spaces would become very large in dealing for example with such animals as elephants), some trait or group of traits will prove crucial in making a correct classification. For Aristotle, as well as for the numerical taxonomist and the cladist, one begins with a description of the phenotype. What is important to both modern taxonomy and Aristotle's taxonomy is that these differentiae not be singly favored. This is because there is the problem of why one should choose one "singly favored" trait above another. Aristotle asks why one should choose "wile and tame"—what make them special? If the last differentia is the sole differentia (Aristotle's first argument 642b 8, cf. 643b 30, Topics 144b3), and there is only one such differentia (clear from the nature of division; if there were more than one, another division would be in order), then Man or any other animal must be that final differentia. The ultimate species should equal the ultimate differentia. But this is absurd for it would define Man as cloven footed only, whereas Man has many differentiae (644a 1-12).

A single differentia cannot put an animal into its natural group because it ignores its total form. Division fails by being randomly restrictive (e.g., placing land and water animals into groups) or by not providing enough information through only offering one differentia.

Aristotle's third objection (642b 20) concerns privatives. Aristotle says that privatives do not count as differentiae. A negative attribute does not point to any positive species. There is no species which is featherless or footless. This part of the argument is not followed by Aristotle himself when he creates the large group of bloodless (anima) creatures. "Bloodless" is a privative and is used as an attribute of one of his large genera (H.A. 490b 7, cf. Bonitz, Ind. Ar. 246a 27).

The second part of the argument states that every differentia must belong uniquely to one species (642b 35). This is true because the differentia is that which picks out the species. Now privative properties may belong to more than one species, e.g., featherless biped may belong to Man and to Chicken (cf. Topics 124a). Thus, privative differentiae are not unique because they point to multiple species.{70}

In the fourth objection Aristotle examines a tenet of dichotomists that it is necessary that the number of differentiae equal the number of species. If this is so, then the dichotomist cannot place an indivisible species under several lines of division (or it will have several differentiae). Likewise, it is impossible to have several species under one division (for then the differentia would not be unique to one species). Thus, of the three methods logically available (illustrated in the figures below) (a) and (b) must be ruled out. (C) must also be impossible since division by opposites implies that the differentiae will be some multiple of two.{71}

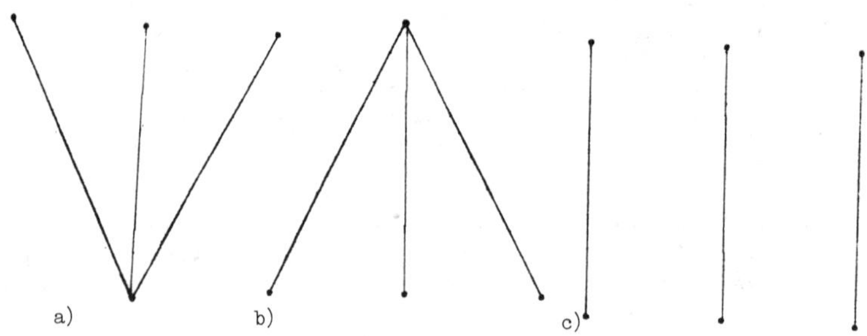

FIGURE TWO: Dichotomy and Division

The fifth argument differentiates between essential and inseparable attributes, a point further developed elsewhere (Metaph. 1025a 14, 1065a 25; cf. Bonitz, Ind. Ar. 714a 20).{72} The last argument is similar to the second. It deals with division by opposites. Many of these divisions will place the same animal into two species. For example, an ant might belong to the "winged" and "wingless" group since at one time in its life it is winged and then at another it is wingless. Likewise, "wild" and "tame" are not proper categories since almost all tame species also have wild members, e.g., dogs (643b 5).

These two examples are said to be the inevitable result of any division in which a single line of differentiation is taken. For such differentiation either breaks up natural groups or it violates its own principles. This happens when one separates animals into "winged" and "wingless" and then divides "winged" into "wild" and "tame." But wild and tame are not really divisions of winged at all. This is adding differentiae to an inappropriate category (643b 20). Either result fails to put animals into the groups that are commonly recognized to exist. The reason for this is because division is a logical exercise which does not recognize the imperative of arranging groups of animals according to the way they actually are (or are commonly believed to be). Aristotle found division to be lacking precisely because of these effects which occur due to dichotomy's *a priori* methodology. One can sum up Aristotle's objections to dichotomy into two kinds. He attempts to show internal inconsistencies in his first, third, fourth and fifth objections, while in objections two and six he examines the results of dichotomy against groups commonly held to be natural classes.

Another way of examining these objections is as exemplifying dichotomy's descriptive and predictive power. Dichotomy cannot adequately describe the natural group because of methodological errors, viz., the use of one differentia which breaks up natural groups etc. These internal flaws make one suspect that an unknown animal would not be classified in a way which would isolate its essence, since dichotomy seems to miss essences (643a 27).

Aristotle's systematics, on the other hand, has explanatory power. Through the dual uses of *eidos*, the working principle of a biological system can be explained since discovering (form) is to explain how a system (*eidos* as species) actually operates, (cf. also his use of *genos*.) Once this is determined one understands the essence of the system (has scientific knowledge of it). This search for the *ti esti* of some form is consistent with Aristotle's more detailed discussion in the *Posterior Analytics* on definition. In 2,3 Aristotle explains that a definition provides the essential nature of some subject. It is an indemonstrable account of essence (94a 11). These essential attributes are found by looking for common traits.

> We must set about our search by looking out for a group of things which are alike in the sense of being specifically indifferent, and by asking what they all have in common; then we must do the same with another group in the same genus and belonging to the same species as one another, but to a species different from that of the first group. When we have discovered in the case of this second group what its members have in common, and similarly in the case of all the other groups, we must consider again whether the common features which we have established have any feature which is common to them all, until we reach a single expression. This will be the required definition. (APo. 97b 7-13).

This searching involved a hunting and enumeration of attributes. "Let us now consider how we should hunt for the attributes which are predicated as elements in the definition" (APo. 96a 22). This hunting requires special rules to guard against the errors of the dichotomists. The enumeration of attributes together cannot have a wider extension than the subject (96a35). When all these conditions are met, then a definition is achieved. Now these attributes refer to shape and function (see the discussion of the dolphin in 4.5). This is emphasized in the Metaphysics where definition is said to relate to the function of some entity, "Each part, if it is to be properly defined, will not be defined apart from its function" (1035b 16). In Chapter 4 it will be shown how Aristotle follows his statements on definition (from the Organon) and taxonomy (from P.A. 1 consistent with the arguments in the APo.; see the discussion in Chapter 5) in his biological practice.

Finally it should be mentioned that as good as Aristotle's classificatory scheme is, it has some real shortcomings when compared to any of the modern schools. Modern standards are clearly superior because they are more exhaustive and have a better way of dealing with difficult cases. For example, any modern system has a way to classify a novel form within its general framework of groups and sub-groups without having to create a special class for it as Aristotle does (e.g., with the whale). Aristotle's methodological motivation for acting as he does may be due to critical empiricism, but by creating intermediate groups he runs the risk of violating the dictums of ontological simplicity which I, along with others, such as Balme (1972) have attributed to him. A proliferation of natural kinds does not fit well with such a view (nor would it fit in with psychological or with linguistic simplicity). However, though Aristotle may fall short as a practioner of alpha/beta taxonomy and identification, these were not his ultimate intention. What he offers is sound insight into the fundamentals of meta-beta taxonomy. His suggestions for constructing definitions and for the structuring of systematics were far superior to any offered by his contemporaries and form a starting point for modern work.

Notes to Chapter Two

1
'Biologist' or 'Zoologist' of course were not terms in use at this time. I will use them, however, in their modern senses to avoid confusion.

2
Aristotle's *De partibus animalium I* translation and commentary by David Balme (Oxford, 1972), p. 70.

3
There is quite an extensive literature on the question of simplicity. For an overview of some of the more important recent contributions see: W.V.O. Quine, "On Simple Theories of a Complex World" *Synthese* 14 (1963): 103- 106; Nelson Goodman, "The Logical Simplicity of Predicates" *Journal of Symbolic Logic* 14 (1949): 32-41; "New Notes on Simplicity" *ibid.* 17 (1952): 198-191, "Axiomatic Measurement of Simplicity," *Journal of Philosophy* 52 (1955): 109-122; Richard Rudner, "An Introduction to Simplicity" *Philosophy of Science* 28 (1961): 109-119; S.F. Barker, *Induction and Hypothesis* (Ithaca, New York, 1957), pp. 91-105; Karl R. Popper, *The Logic of Scientific Discovery* (N.Y., 1959), 6, 7, 9; H. Reichenbach, *Experience and Prediction* (Chicago, 1938), section 42; F. Attneave and R. Frost, "The Determination of Perceived Tridemensional Orientation by Minimum Criteria" *Perception and Psychophysics* 6.6B (1969): 391-6; S.W. Blackburn, "Goodman's Paradox," in N. Reshcher ed., *Studies in the Philosophy of Science, American Philosophical Quarterly Publication* no. 3 (1969): 128-42; H. Putnam, "Reductionism and the Nature of Psychology" *Cognition* 2.1 (1973): 131-46; W.C. Salmon, "Statistical Explanation" in W.C. Salmon ed., *Statistical Explanation and Statistical Relevance* (Pittsburgh, 1971), pp. 29-88. Eliot Sober, *Simplicity* (Oxford, 1975).

4
Aristotle sees this point as being in open debate (639b 8).

5
Balme, 1972, p. 73.

6
Though my reading of 639b 9–11 follows the standard rendering of these lines an interesting alternative was recently suggested to me. In this alternate interpretation (contradictory to mine) the order of emphasis changes. In the standard rendering one takes _prōton . . . epeith'_ as "first . . . (the data) . . . and only _then_ (the theories) . . ." However, this alternate reading suggests that Aristotle might consider empiricism to be understood. That is, _epeith'_ may simply mean "thereupon"; and the clause which follows may be for the real purpose of introducing the idea of causes, so that the contrast with "another procedure" may have in mind one which does not deal with causes at all and is "uncritically empiricist."

7
This repetition of the dictum of psychological or linguistic simplicity gives some support to those who wish to attribute such a position to Aristotle.

8
Aristotle's lack of a technical vocabulary is discussed by Maurice Manquat, _Aristote Naturaliste_ (Paris, 1932), pp. 104–107.

9
Aristotle used _genos_ and _eidos_ technically as genus and species, but he had no name for particular kinds.

10
Examples of difficult cases are snakes (H.A. 505b 5) and sponges (H.A. 548a 33).

11
It should be remembered that in the somewhat primitive Greek Society, beekeepers and the like were considered experts in their respective areas. They were the experts who do not explore causes (mentioned at 639a). Thus, the reliance on "common opinion," here, is equivalent to examination of "expert's testimony."

12
See 2.4 for further explication of this term.

13
Cf. *Nicomachean Ethics* 1, 8; 7, 1; 10, 2.

14
Aristotle does not make this mistake (H.A. 489b 2). The point is that popular groups cannot be accepted carte blanche. One may accept that there is a group called Birds or Fishes, but just exactly how many and what kind of species will be put into the genus must be the subject of the biologist.

15
Aristotle is showing his awareness of the restrictions of common language here.

16
This is not the only place that Aristotle speaks of the joys of scientific knowledge cf. Metaph. 980a 1, 1048b 18; and that theoretical activity is an essential element to *eudaimonia* E.N. 1177a 12, b 29; 1178b 8, 21.

17
Aristotle's example of Heracleitus in the kitchen (645a 21) implies that Aristotle liked to get into the "field" and do biological research. This is a little different from the picture of the "arm chair" biologist painted by Duering, "Aristotle's Method in Biology" *Aristote et les Problems de Methode* (Louvain, 1961), pp. 213-221.

18
David Balme, "GENOS and EIDOS in Aristotle's Biology" *Classical Quarterly* 12 (1962): 81-98; A.C. Lloyd, "Genus, Species and Ordered Series in Aristotle" *Phronesis* 7 (1962): 67-90; Richard Rorty, "Genus as Matter" in *Exegesis and Argument* Phronesis supplement (Assen, Netherlands, 1973), pp. 393-420; Marjorie Grene, "Is Genus to Species as Matter to Form?" *Synthese* 28 (1974): 51-69; Anthony Preus, "EIDOS as Norm in Aristotle's Biology" *Nature and System* 1 (1979): 79-103; Jim Lennox, "Aristotle on Genera, Species and 'the More and the Less'" *Journal of the History of Biology* 13.2 (1980): 321-346.

19
Cf. David Balme, "Aristotle's Use of Differentiae in Zoology" *Aristote et les Problèms de Méthod* (Louvain, 1961), pp. 195-212.

20
Aristotle does not explicate (d) here, even though he explicates a-c. For this quotation I am following Trendennick's Loeb translation.

21
Other passages include Metaph. 1058a 22, b 22-26; 1045a 23.

22
See Marjorie Grene, 1974.

23
Thompson seems to think that these lines are a technical phrase of some sort cf. _De caelo_ 312a 12, _Physics_ 207b 1. Balme disagrees, citing the logical usage in _Topics_ 121b 25, 140a 2, 144a 12.

24
Topics 101b 17; 102a 31, b 2;, 103a 8, b,5.

25
From Ross's note in the Oxford Translation.

26
P.A. 644b 10, Metaph. 1016b 33. By "matter" here it is meant the shape and construction of the _homoiomerē_ _and_ _the_ _anomoiomerē_.

27
For Aristotle's examples, P.A. 644b 10, 645b 5.

28
Cf. G.A. 761b 10.

29
Lennox, P. 336.

30
It should be noted that Lennox seems to deny that _genos_ and _eidos_ refer to class concepts, p. 336.

31
Since Aristotle has no theory of evolution, these general classes must be derived logically or temporally through generation and ontogenesis. This constitutes what I will call a quasi-phylogeny. There is no real ancestry except in Aristotle's queer account of the development of the fetus in

which plant, animal and various subtle gradations are passed through before the fetus becomes human.

32
The central work of this school is by Willi Hennig, Phylogenetic Systematics tr. D. Dwight Davis and Rainer Zangerl (Urbana, Illinois, 1966) revision of Grundzuege einer Theorie der phylogenetischen Systematik (1950).

33
Charles Singer, "Greek Biology and its Relation to the Rise of Modern Biology" Studies in the History and Method of Science ed., Charles Singer, vol. 2 (Oxford, 1921), pp. 20-22.

34
These have been done by Norman Platnick and G. Nelson in Systematics and Biogeography: Cladistics and Vicariance (N.Y., 1981), pp. 69, 71, 73.

35
Jurgen Meyer, Aristotleles Thierkunde (Berlin, 1855); Aubert and Wimmer, Aristoteles Thierkunde (Berlin, 1868), pp. 58-184.

36
The additions to the list given in brackets are mine.

37
Others have made Aristotle out to have been interested in establishing a complete classification system. Meyer (p. 229) claims this leads to absurdities while others like Harold Hantz, The Biological Motivation in Aristotle (N.Y., 1939), pp. 109-114, claim success for Aristotle in this endeavor.

38
Singer, pp. 10-11. A.W.H. Adkins claims the same is true of modern Greeks.

39
There are several possible meanings to "complete." It is not the case that Aristotle did not have some system into which he could fit any animal, but rather that the system he does offer is not satisfactory for the reasons given above.

40
No coherent classification method comes out of the Hippocratic writings.

41
G.G. Simpson, The Principles of Animal Taxonomy (N.Y., 1961).

42
Ernst Mayr, Principles of Systematic Zoology (N.Y., 1969).

43
J.R. Gregg, The Language of Taxonomy (N.Y., 1954).

44
See Simpson (1961) and "The Principles of Classification and a Classification of Mammals" Bull. Amer. Mus. Nat. Hist. 85 (1945): 1-350; Mary (1969) and D.H. Colless, " A Cornucopeia of Categories" Systematic Zoology 26 (1977): 349-352.

45
R.R. Sokal and P.H.A. Sneath, Principles of Numerical Taxonomy (San Francisco, 1963). Numerical Taxonomy (San Francisco, 1973).

46
Besides those of Hennig, other important works in cladistics include: G. Nelson, "Cladistic Analysis and Synthesis: Principles and Definitions" Systematic Zoology 28.1 (1979): 1-21, "From Candolle to Croizat: Comments on the History of Biogeography" Journal of the History of Biology 11.2 (1978): 269-305; N. Platnick and G. Nelson, "A Method of Analysis for Historical Biogeography" Systematic Zoology 27.1 (1978): 1-16; Nelson and Platnick, op.cit., 1981; E.S. Gaffney, "An Introduction to the Logic of Phylogeny Reconstruction" in Phylogenetic Analysis and Paleontology J. Cracraft and N. Eldredge eds., (N.Y., 1979), pp. 79-111; N. Eldredge and J. Cracraft, Phylogenetic Patterns and the Evolutionary Process (N.Y., 1980).

47
Niles and Eldredge, 1980, p. 10.

48
Under this system Mammal would become an order while Marsupialia and Placentalia would be downgraded to families and the orders of Placentalia would be tribes (Hennig, 1966, p. 187).

49
E. Mayr, "Numerical Phenetics and Taxonomy Theory," _Systematic Zoology_ 14 (1965): 75.

50
C. Linnaeus, _Systema Naturae_ 10th ed. (Stockholm, 1758).

51
In a slightly different way, Michael Rohr makes a comparison between Aristotle's use of classification in the _Topics_ and cladism in his paper "Aristotle and Genus," presented at the Conference on Aristotle's Biology, Marquette University, 1981.

52
This is a tricky point, for since Simpson (1961, p. 78) links phylogeny to homology, which he said in "Anatomy and Morphology: Classification and Evolution: 1859-1959: _Proc. Amer. Phil. Society_ 103 (1959): 286-306, is concerned with similarity. This is echoed by W.J. Bock, "Foundations and Methods of Evolutionary Classification" in M.K. Hecht, P.C. Goody and B.M. Hecht eds. _Major Patterns in Vertebrate Evolution_ (N.Y., 1977), pp. 851-895, claiming that "shared similarity is the _only_ valid empirical test of homology" (p. 882). Though Sokal and Sneath would demur since they do not wish to recreate phylogenies with homology, it is this sense that I am appealing to in my expression "quasi-phylogeny." This seems to be the concern of both the cladists and the evolutionary biologists. On the other hand, like the pheneticists, Aristotle is groping for a method to weigh physical appearances and the way in which these similarities actually contribute to specific similarity.

53
See Lennox's discussion, p. 337ff.

54
These lines can also be read literally as Peck translates them, "Footed, two-footed, cloven footed." The two readings are consistent with my interpretation, though the aspect of repetition is emphasized by the rendering given in my text.

55
Aristotle might be critizing Plato in the _Sophist_, or as J.M. Le Blonde, _Aristote: Philosophe de la Vie_ (Aubier, 1945), n. 93, suggests, Speusippus. I believe this is just a general reference to an unpublished discussion in the academy.

56
The reader may note here some parallels here which have been found with Saul Kripke, "Naming and Necessity" in _Semantics and Natural Language_, G. Harman and D. Davidson eds. (Dordrecht, 1972), pp. 253-355, cf. Hilary Putnam, "Dreaming and Depth Grammar" in _Analytical Philosophy_ ser. 1, ed. R.J. Butler (Oxford, 1962); "The Meaning of 'Meaning'" in _Minnesota Studies in the Philosophy of Science_ 7, ed. K. Gunderson (Minneapolis, 1975). Kripke and Putnam state (along with Aristotle) that our words for 'kinds' make illusion to an unspecified essence. However, the manner in which such a reference is fixed via "rigid designation" differs from Aristotle. For an interesting comparison between these philosophers, see R. Sorabji, _Necessity, Cause and Blame_ (Ithaca, N.Y., 1980), pp. 210-211, 219-23.

57
This claim is well documented by A. Cain, "Logic and Memory in Linnaeus' System of Taxonomy" _Proc. of the Linnaeus Society_ 169 (1958): 144-163; H.K. Svenson, "On the Descriptive Method of Linnaeus" _Rhodora_ 47 (1945): 363-388; C.H. Lindroth, "The Linnaean Species of Carabid Beetles" _J. Linnaeus Society Zoology_ 43 (1957): 325-341.

58
Plato probably would be a better candidate here. Aristotle's essentialism has been attacked in this respect by Mayr, (1969), M.T. Ghiselin, _The Triumph of the Darwinian Method_ (Berkeley, 1969) and David Hull, "The Effect of Essentialism on Taxonomy--Two Thousand Years of Stasis" _British Journal of the Philosophy of Science_ 15 (1965): 314-326, and "Why Species Cannot Evolve" presented to the Conference on Aristotle's Biology, Marquette University, 1981. For opposing views see: Preus (1979); Lennox (1980); Norman Platnick and Donn E. Rosen, "Popper and Evolutionary Novelties" forthcoming volume on Popper ed. Paul Levinson; David Balme, "Aristotle's Biology was not Essentialist" _Archiv fuer Geschichte der Philosophie_ 62.1 (1980): 1-12.

59
See Preus (1979) and Montgomery Furth's "Notes on the Construction of Aristotelian Biological Individuals," presented to the Conference on Aristotle's Biology, Princeton University, 1976.

60
For further discussion of this point see A.L. Peck on "dualizers" in his introduction to the Loeb *Historia animalium*, sections 28-33. See also Anthony Preus, *Science and Philosophy in Aristotle's Biological Works* (N.Y., 1975), pp. 158ff.

61
Marjorie Grene, "Aristotle and Modern Biology" *Journal of the History of Ideas* 33 (1972): 395-424.

62
It is interesting to note a similar point made by D. Johanson and M. Eddy in *Lucy* (N.Y., 1981) in discussing a biped primate, which is supposedly an ancestor to *Homo Sapiens*. Whereas I do not agree with much of their presentation, it is curious that they make the same Aristotelian argument about Lucy's upright posture (cf. P.A. 1 on segmented spines).

63
G. Cuvier, *Le Regne Animal Distribue après son Organisation*, 2 ed. (Paris, 1835). For moderns, this is true because of phylogeny. For Aristotle, it was true because *logos* is prior to nature, and the genera is the *logos* of the species (P.A. 640a 17).

64
It is important to note that 'why' does not mean "How does it come to have," but refers to essential predication on various levels of description, viz., the levels of species and genus.

65
Darwin, *Orgin of Species* (London, 1859), ch. 15; Paul R. Ehrlich and Richard Holm, "Patterns and Populations" *Science* 137 (1962): 652-57.

66
Even if these arguments are only directed at an individual who practiced dichotomy they still attack dichotomy and so are not merely *ad hominem*.

67
Wolfgang Wieland, *Die Aristotelische Physik* (Gottingen, 1962), 94n. argues that even modern physicists interpret experiments in an Aristotelian manner, "Auch die neuzeitliche Physik liest an einem speziellen experimentellen Fall eine allgemeine Regel ab und pruft dann, ob es sich um die wahre Allgemeinheit handelt. Erst dann namlich kann sie sehen, ob sie

von einem speziellen Fall ausgegangen ist oder nicht; d.h. ob sie diesen Fall von seinen besonderen oder von seinen allgemeinen Merkmalen her gedeutet hat." It is modern theories of induction that have misled us.

68
See the discussion of eidos in 2.3.

69
This is not the case in Numerical Taxonomy, see 2.4.

70
"Blooded" in 643a 6 is a common characteristic shared by Man and Bird, but it is shared in a different way than "biped" is shared. Aristotle wants to show here that these larger groups, though they may not be commonly put together, share a tie of sorts. But by asserting that they have "different blood," he is also showing that common attributes of a larger group are not as binding as those traits shared by genera.

71
P.A. 643a 12ff.

72
Aristotle's attitude toward 'properties,' 'essential attributes' and 'accidents' has changed here, according to Le Blonde. Poion in Topics 122b 12-17, 144a 20, "La différence, qui est dans l'essence, détermine le poionde l'essence, mais elle n'est pas cependant le poion; elle est essence. En fait, et dans ce livre même, Aristote reconnaît qu'a l'intérieur des genres les espèces ne diffèrent que par des somatika pathē qui se ramènent à des différences de grandeurs (cf. 644b 13; 644a 17). On peut se demander dans quelle measure ces pathē appartiennent vraiment plûtot à l'essence qu'au poion" (note 111).

CHAPTER 3

THE METHOD

3.1 THE TWO MODES INTRODUCED

Before being able to proceed to the biological works themselves, one last important area of methodology must be examined. This is the nature of and relationship between teleology and necessity. These are introduced by Aristotle at *Physics* 198b.

> We must consider: 1) Why Nature acts among the causes for-the-sake-of-which, and 2) How necessity works in the sphere of nature/Nature. For all writers refer to the necessary cause. They say that since hot and cold have such and such a nature, these come to be out of necessity.

Again at P.A. 642a 2: "There are then two causes: 1) That for-the-sake-of- which, and 2) That from necessity." Now the seasoned Aristotelian might, at first, want to describe all physical events (events in *phusis*) in terms of the four causes. This would not be a bad beginning, except that Aristotle seems to be talking here about two modes of explanation and not four. The reason for this is that in biology (as well as, perhaps, other branches of natural philosophy) there is some grouping of the four causes. Corresponding to purpose (*heneka tou*) are the formal and final causes. This link is made explicit at *De generatione animalium* 715a 9, "The formal cause and the final cause (when it is considered as an end) are the same." This explicit link is implicitly supported at P.A. 639b 14, "Clearly the final cause is first [compared to the efficient cause] because it is the definition, and the definition is the ground both for the products of Art as well as Nature." In these two passages the final cause and the formal cause (which as we saw in Chapter One stands for the essence as well) are linked. In the P.A. passage final cause is linked to the *logos* (definition/essence) just as form was linked to the *to ti ēn einai* in Metaphysics 1032b 15. These four concepts are *present in the* productive sciences.{1} In the last chapter the various senses of *eidos* were discussed (2.3). In this case *eidos* is close in *meaning* to the essences discussed in 2.5 with *reference* to Aristotle's systematics.

Thus we see that the formal and final causes are united in biological explanation. They go together to form a strengthened form of teleological explanation (TE) which will cover the normal final cause as well as *eidos* in its various forms and associated essence terms such as *logos*, *ti esti*, and *to ti ēn einai*.

The second half of this chapter will deal with necessity. The grouping here is perhaps more difficult, but it revolves around the close association which is made between necessity and the material cause. At Phys. 200a 15 Aristotle says that "Necessity is in the matter while that-for-the-sake-of-which is in the definition (*logos*)." Here again the dual system of explanation is maintained. Necessity is linked to the material cause, i.e., matter and its motions. Since matter *acts and reacts*, both through its own constitution (e.g., the strength of *bricks* can contribute to the strength of a wall, cf. 200a 5), and its motions, the efficient cause must be brought in here. When describing necessity through the motions of matter, Aristotle describes a dynamic event.

> Now those in ancient times who were the first to describe nature, talked about material origins . . . i.e., what kind of thing is matter and how the universe is formed out of it and what caused movement . . . this assumes that the underlying material had a nature out of necessity, e.g., fire is hot, earth cold.(P.A. 640b 4-10).

In this passage the motive and material elements are combined. For Aristotle did not think of matter only in terms of its passive, structural properties, but included its associated motions. This creates a kind of quasi-mechanical explanation of how events come-to-be. This material (or quasi-mechanical) explanation will be termed ME and will cover those events governed by necessity. It will include both the material *and* the efficient causes. Thus, Aristotle's explanatory framework in his biology is seen to accord with his modes of explanation elsewhere. To speak of events happening due to a purpose and necessity, then, does not violate the doctrine of the four causes, but merely makes a convenient grouping for discussing certain natural events. Because these two modes are so important in understanding Aristotle's biological method each will be examined separately in the succeeding sections in order to complete our outline of Aristotelian biological method.

3.2 TELEOLOGY: ART AND NATURE

At the end of the 639b 14 passage quoted in the last section Aristotle compares the way in which final cause works in Nature to the way in which it can work in Art. These parallels are developed further.

> For we define conceptually and perceptually, as the doctor does with health on the one hand, and the builder does with his house on the other. After the logos is acquired, each can expound the causes and the reasons why. But the final cause is better and fairer in Nature's work than in the artisan's. (P.A. 639b 17-20).

Now it is very easy to imagine final cause as operative in art or in building. Presumably the painter or builder has a plan in mind before he begins painting or building. The paint is applied to the canvas according to a design and color scheme which the painter already has in mind. But how does one apply this to Nature? One could assume that Nature is like the painter, i.e., consciously creating its objects. There are several passages in which Aristotle seems to speak this way, e.g., Nature aims at the Best and Noble, P.A. 639b 20, 645a 25, 670b 24; Nature acts Rationally, G.A. 731a 24, 744b 15; Nature Compensates, G.A. 760b 26. Though the language employs the metaphor of an active creator, one is inclined to treat it only as a metaphor since Aristotle nowhere accounts for any active intelligence in Nature.{2} If there were such a force, one would expect a comprehensive account of it. Lacking this one may suppose that the usage is merely metaphorical.

But this is not sharp enough. In order to understand exactly what Nature is, a few crucial texts must be examined.{3}

One group of texts suggests Nature is an overarching, regulative force.

Nature does nothing in vain. P.A. 658a 9, cf. D.C. 271a 33.

In all its workings Nature acts rationally. G.A. 731a 24, cf. 739b 20, 41b 1, 44a 35.

Since Nature has taken away from their numbers it compensates in giving them greater size. G.A. 760b 28, cf. 750a 5; P.A. 657b 38, 695a 11, 674b 30; I.A. 712b 18; I.V. 691b 9.

Nature distributes to each what it can use. P.A. 687a 11, cf. 655a 27.

Nature uses a part for two functions. De An. 420b 18, P.A. 659b 34, 662a 16, 683a 20.

(See also the comparisons with art, G.A. 734b 20, 735a 1, 762a 15, 775a 20 and with a craftsman G.A. 731a 20, 730b 15, 743a 25.)

Another group emphasizes individual reproductive natures.

It is through this that generation creates another like itself. For all plants and animals have this function in their nature. G.A. 735a 18.

The most natural event is the generation of another like itself: an animal producing an animal, a plant a plant, for the purpose (according to nature) of partaking in the eternal and the divine. (De An. 414 28).

A third group introduces the <u>eidos</u> or species (cf. 2.3).

It is through these causes that there is generation of animals. Since it is impossible that the nature of this type be eternal, it is according to another way that these are eternal. It is impossible that this be numerically, for the being of things is in the particular and if it were so it would be eternal. However it is possible via the species. G.A. 731b 32-35.

Since no living thing can participate in the divine . . . it achieves what is open to it . . . it does not remain the same individual but continues in that which is like itself--not numerically but specifically one. De An. 415b 7, cf. G.C. 338b 13.

> Both the individual and the genus are present in generation, but the individual is more important because it is in him that being resides. G.A. 767b 34.

What does one make of these passages? It might seem that on the one hand Nature is an overarching regulative agent, while on the other, it is nothing more than individuals transmitting their own characteristics. In typical Aristotelian fashion the correct answer lies in both approaches. Nature is an overarching entity. It can redistribute and create an ecological balance as the first group suggests.{4} But it is not an actual persona. It is a principle. This principle is simply that the fittest survive.{5} They do survive and the characteristics of the classes are molded by this single fact. Thus this principle acts as a real cause. It does regulate the form and activity of organisms. It is this principle (in conjunction with other physical principles) which instills regularity among the species and allows the activity of biological classification. This observed regularity demonstrates a controlling force— this force is Nature (cf. APo. 95a 9, P.A. 640a 20; also see Chapter Five). As a real cause it is no mere 'als—ob' Reflexionsbegriff as some have claimed.{6} It acts as a constraint upon the development and continuance of an individual and *a fortirori* of a species. One consequence of this principle is the exitence of natural kinds—groups (the species) which exhibit common, family resemblence. In fact when this regulative principle does not master the *katamēnia* of some individual event of genesis, the result is a "*relapse*" (a movement away from the parent as well as away from the species) which can cause deformities, G.A. 4.3.

But at the same time it is also true that individual natures are expressed *only* from individual to individual. This is the emphasis of the group two passages. What makes Socrates or Coriscus unique is a set of accidents passed on by their parents and uniquely occurring in them, G.A. 767b 32, cf. Metaph. 1015b 17. These manifestations of individuality are from the "nature" of the progenitors. If the male's *sperma* moves in such and such a way, then the resultant nose may have a bump or a "snubbed" cavity. All individual traits are due to the movements of the *sperma* (768a) and *katamēnia* (G.A. 4.4). Thus, the unique way that some individual is constituted may also be termed 'nature.'

With two senses of the word there may be some confusion, however; the third set of quotations offers a good perspective. Whereas individual nature is important to the thing being generated, (because it is precisely those individual traits that make him a unique, actual person), to the biologist, the larger group is more important. This is because the larger group, the species, persists through time. In order for the biologist's investigations to have any general scope at all, he must make the species his principal object of investigation (it is only the sophist who studies accidents, Metaph. 1026b 15, cf. Soph.Elench. 173a 34). For it is the species which exhibits essence (1030a 12) and since true definition is only of the essence (1030b 5), one may conclude that only species may be defined and investigated *per se*. Thus the biologist properly concentrates upon species and upon specific characteristics. (This is, in fact, the way Aristotle properly proceeds in the *De partibus animalium*, both in theory, 639a 16ff., and in practice, Books 2-4).

So even though to Socrates, his accidents, which have been passed on to him, are of great personal interest qua individual (because without them he would not be the particular man, Socrates), still to the biologist, such a nature is of less interest than the nature of Man. It is this nature which is eternal{7} and more divine. It is this nature which can reveal biological truth.

Both the individual natures of various men and the nature of Man are in turn subject to the principle of fitness. The acquisition and loss of parts— even the formation of accidents, properties or essential attributes may be affected by this law-like reality. Aristotle often describes some animal having or not having a part due to the part's contribution to that species' fitness (P.A. 662b 1-16, 674a 28, 683b 18, 693a 3-19, 694b 6 *et passim*). Thus, in addition to the natures of x (where x is an individual or species), there is also the Natural regulating principle of x which regulates all possible substitution instances. This Natural regulating principle is the fitness dictum. It is this dictum which is the overarching principle alluded to in the first group of passages and governs the natures of the latter two groups. Thus 'nature' can refer to either depending upon the context. [For purposes of clarification, I have capitalized 'Nature' when it refers to the overarching principle.]

But even at this there are some possible difficulties with teleology. It is perfectly understandable in the case of a painter or a builder when it is asserted that the end of a process affects its generation. However, in the case of the development of a man, the case is, at first glance, fraught with difficulties concerning causation. What would it mean to say the essence of a man affects his generation? One could see this as the man when finished (adult) affects the generation and growth of all preceding stages. The man as adult at t affects the child's development at (t - n). This appears to be making the future affect the present.{8} If x is presently a child, then is x to understand that his development is being determined by what he is going to be? The future, by definition, is that which is not yet. And we know from Aristotle's famous discussion of the sea battle, the future is not yet determined.{9} Therefore, it seems impossible for that which is not (at present, and therefore is undetermined with respect to the adult), to in any way affect that which comes before it. In this sense, x as a child is not affected by x as x will be as an adult.

To make sense of Aristotle's teleology, the effect of the end upon its development must be understood in another way. There are at least two ways in which this comparison makes sense. The first is exemplified in the modern conception of DNA and its effect upon human development.{10} It is accepted that virtually all phenotypic traits result from processes controlled (at least originally) from the ordering of the four bases: adenine, guanine, thymine and cytosine. This can be demonstrated by examining several models of genotypic systems. To be determined by an end (i.e., to be consistent with Aristotle's teleology) one would expect that certain structural factors, innate input (to be called I) determine certain ends which result in phenotypic characteristics. Against this would be the so-called "chance" factors, those contributing inputs which are not structurally coded and may even alter certain set structural arrangements (i.e., intra-system environmental factors, E). These two effects will be judged within some system, A. The first view to be discussed involves simple transcription and translation for the amino acid, phenylalanine. If we view A as being the triplet sequence UUU (three uracil nucleotides in a row), then A is able to transcribe this information onto a complementary RNA chain which in turn translates this into a protein, given that the messenger RNA molecule consists in at least 300 nucleotides (translating into a protein of 100 amino acids). The description domain of this example will be just this process at this level. It has been assumed that A can code for phenylalanine. The action (y) with regard to some characteristics (z) is seen to be controlled by A and its structural qualities (x), though it has to be admitted that environmental factors are important also in beginning and carrying out this process.{11} What is clear, though, from this example is that on this level and under this description, A can be said to perform y with regards to z due primarily to A(x)—in this case, A's physical structure. Given this account, then A's "actions" can be said to be primarily teleological (due to I, see appendix note 3).

This characterization of the genetic level becomes more difficult, however, when we consider the second example of position effect. In the first example, it was the structure of A which determined y. There are other examples where this causal interaction is not so clear. If we consider a "gene" in this example to be one cistron long and to have a particular structure designed for a single function (as in the first example), then this model is an example of I which is inconsistent with work done early in this century on <u>drosophila.</u>{12} This work indicated that the position of a gene affects its expression. For example, one gene structure, x, in A produces an evenly distributed eye pigment in one position, and in another position A(x) produces a variegated pigment (position was determined by staining techniques). In this example we have A(x) => y at locus 1 and A(x) => y' at locus 2. In both instances A is the same structurally, but its effects change, therefore; it is incorrect to make y a simple effect of A(x) when locus is so important in y's composition (expression of z through A(x)). The locus in this example can be viewed as the environment. E affects A within a range of two choices. A is responsible for there being two and only two choices, but regarding the expression of those choices, it is dependent upon E.

In these first two examples the distinctions between A and its environment have become somewhat clear, but in the third example, the rélativity of such judgments becomes evident. In cases where multiple genes affect the same trait, another complication, epistasis, may be introduced, whereby one gene affects another. For example, cross two mice, one homozygous for the dominant color, gray, and the other homozygous for the recessive color, black. One would expect that all the offspring would be gray, but occasionally there is an albino (at a rate too high for mutation). The explanation concerns a second set of alleles at another locus which also affects color. If there is a dominant allele at this locus, then the decision about coat color is made at the primary locus. The recessive allele at this secondary locus is for albinism so that any mouse with two recessive alleles at the secondary locus will be albino irregardless of the coding at the primary locus. This second gene is said to be epistatic to the first because there is a sense in which it regulates the first. Now, the problem for our analysis concerns which to consider as A: the primary or the secondary locus. If one were to consider simply the desired effect y, then it would seem that either would do equally well within certain circumstances. If the structure of the primary allele is taken as A, then structure is being opted over function as the point of reference, since there are instances where the structure of the primry allele is irrelevant to the coat color of the mice. The advantage of this latter choice is that it allows a fixed definition of A under the description that the primary allele will be viewed as A(x). If one wanted to alter the description, it would be possible to make the structure of the allelle at the secondary locus the central point of reference, but then it would be cogent to a different description.

If the structure of the allele at the primary locus is chosen as A, then it is clear that A => y, just in case the secondary locus is not double recessive. In the instances where it is, then y is solely a resultant of E. Unlike the second example where A was always I (though E picked part of the available choices in which A could be expressed), in this third example there is a case where A is unrelated to y as a primary efficient or final cause, and the effect may be viewed as totally environmentally determined.

Position effect, epistasis and other factors, such as operon theory, all alter the basic teleology due to structure. In this last example, the activity of the genome is controlled in a series of steps.{13} Each cell's activity depends on one enzyme or a group of enzymes. Such enzymes are synthesized on ribosomes according to a pattern determined by mRNA (messenger RNA). The mRNA, in turn, is produced by a cistron of DNA. Whether the cistron is working or not is determined by a neighboring operator gene. The operator gene, in turn, is controlled by a regulating substance reaching it from the protoplasm. This regulator substance is produced by a regulator gene located anywhere in the DNA of the cell. Environmental factors such as food substrate, embryonic inducers, hormones etc., can then act on the cell, interacting with the regulating substance in the protoplasm, and can thereby control the activity of the genome. Now, it does not matter which body one wishes to title A (whether it is the cistron of DNA or the operator gene). Each is controlled by a factor external to itself and therefore E, by definition. Any action of A, though mechanically proceeding from A, was caused by E which determines whether or not A will operate at all. Thus E either completely or partially determines y, while the best A can do is partially cause y in some cases. In this final example, E has risen to the dominant factor, when in the first example I was the primary cause of y.

Thus it can be seen that structural and non-structural efficient causes affect the phenotype. The structural causes have been further linked with final causes so that one might wonder whether Aristotle would be comfortable with this explanation or not. However, it should be clear that these various instances of indeterminate genotypic expression deal with traits which Aristotle would consider to be non-essential. From what is presently known, those traits which are essential are structurally controlled by information coded through DNA base sequences. Variation in essential properties is due to point mutation or chromosome aberrations, which are specific interruptions in the genetic code.{14} Now one would expect Aristotle to support modern genetics with respect to its teleological orientation (adopted by many, but not all of its practitioners). It is not the case that some embryo is influenced by itself as it will be as an adult, because that would make the future affect the past. Rather, some embryo is affected by the essence of Man which acts as a regulative end to its own development. This essence is expressed through the species' genetic phylogent representing the species' membership to various genetic populations. The particular "essence" as represented in that population (plus the accidental traits of sub-populations) is a limiting case which controls ontogenetic alterations. It is Man, in general as a species, that is the end of any particular embryo (G.A. 723b, 716a 30, cf. 730b 13). In this way Aristotle need not concern himself with the supposed problems caused by non-structural determinants in the developmental process. Since these non-structural or environmental causes only determine non-essential attributes such as hair color, albinism and the like, they need not have bothered Aristotle. In fact, Aristotle's general account of mutation could fit in here in a very general way.

> The cause of monstrosities occurs when the <u>sperma's</u> motion is disturbed through the matter not being <u>mastered</u> leaving the matter most general--this generality is merely "animal." It is said that the result has the head of a ram or ox and in others similarly, e.g., a calf with a child's head or a sheep with an ox's head. All of these occurances result from these causes I've talked about. (G.A. 769b 10-15, cf. 768b 30, 769b 10).

Aristotle offered a genetic theory in which the eidos of a man was present at conception.

> . . . when the male seed enters into the female, it takes (the katamēnia, menses) and combines [with the katamēnia] into the proper shape. (G.A. 733b 20).

Though vague, this account describes a process in which the essential form of Man is present at fertilization as well as those individuating qualities of the parents (hair color etc., cf. G.A. 778a 33). With the eidos coming from Nature through the male, the essential qualities are physically transmitted by the parents to the offspring. The important point here is that there are various levels of generality operating. First there are the individual traits being transmitted which will make this future adult unique (a mixture of male and female elements, G.A. 715a 20). These traits cannot be seen completely teleologically in themselves since they may be determined by environmental factors along different stages of development (e.g., eye color, 778a 33).{15} The more significant teleological perspective is the continuation of the abstract form of Man (cf. the discussion in 2.5). These are the specific traits which, by definition, are essential because they make the organism a Man. They constitute the ti esti. Such essential attributes can be viewed teleologically because they are in the logos of Man and thus common to all members of the species. In modern theory these traits are embedded in a number of structural DNA configurations common to all men. At a higher level, men are all animals (639a 15, cf. 2.2). These common functions are also illustrated structurally in modern theory as well as in certain structures common to all eucaryotes.{16}

There seems to be a pattern by which organisms develop. Without such a pattern, one would have to assert that the development of Man occurs by chance.

> Therefore, Empedocles was not correct when he said of many animal attributes that they are the result of accidents in development, for example that the vertebrae are accounted for by the fetus twisting so that the backbone is broken. He was ignorant of: 1) That the sperma, the synthetic agent, has a specific power, and 2) The producer was prior not only in logos, but in time as well. (P.A. 640a 20-26).

It is not the case that each backbone is twisted like a stick and broken into many pieces. Such a process would yield great irregularity and variation, but this is not observed to be the case. There could be no defining traits for the eidos if all happened by chance since nothing happened heneka anthropou. "For the sake of" requires a pattern or end to which the thing in question strives. One sense of teleology, therefore, concerns how things actually develop. Do they merely occur by chance with one event accidentally following another or are events connected through a defined process? Aristotle chooses the latter as the only way that complexity in an organism's development can be accounted for.

Another way final cause can be understood is as a necessary tool for any biological explanation. For the biologist to be able to understand the various stages of a chick embryo's development he must have some idea what this embryo will develop into. If the biologist lacked this and did not know whether he was looking at a developing duckbill platypus or a chicken, there would be no way that he could judge the significance of various changes. Indeed, his entire ability to observe anything in detail would be undermined since he would not know what he was looking for.{17} It is essential that the biologist know that this egg is from a chicken. His whole conception of what he sees is largely determined by the telos which he matches against the thing in question.{18}

One way this can be approached is through an examination of the "powers" or "potentialities" (dunameis) located at some basic level within the organism.{19} The modern analogue to this in DNA has already been suggested. For Aristotle the lowest level explanation dealt with a discussion of the contraries: hot, cold, moist and dry. Now it is important to be able to state these properties and discuss them by themselves. (Aristotle does this in the Meterologica and in De generatione et corruptione.) They can be understood through themselves, as well as through a wide variety of effects which they can produce. This is necessary to maintain the explanatory power of these lower level terms. For without this added dimension, the dunamis' character would be hopelessly tied up with the phenomena which it is trying to explain. In such a case it would not be sufficiently independent and could lead one to say that "a capacity for F" explains F (which everyone would admit is rather fishy). Since Aristotle does have a physiological basis for his biological works, independent of his effects, he is not subject to such charges (though he has been incorrectly maligned for this very point).{20}

The power of teleological explanation for modern biology has been called into question by Marjorie Grene among others (see Chapter Five for a more complete discussion of modern discussions of teleology and functionalism).{21} Grene cites W.B.D. Kettlewell's famous experiment on industrial melanism to prove her point. In this study it was found that in a given region there were two kinds of moths: peppered winged and black winged (the mutant of the peppered species). The species live in an area which is subject to industrial pollution. When the trees in this area are blackened by the pollution, the predators of the moths can more easily pick out the peppered winged moths and as a consequence, their numbers decrease. The black winged moths thrive and become dominant in numbers. When the trees are washed, then over a period of months the reverse happens. Grene comments on this:

> Evolutionary speaking, that is what the whole business of being a moth is *for*, not indeed, just "for" the survival of this gene, but if we had a complete count of all the genes in the population at time t-zero and time t-one, the differential ratio would give us the "end" of the story; the differential survival of some genes rather than others. But which genes? Whichever ones survive, of course . . . biological process is first and last and always evolution; evolution is first and last and always a chronicle of survival, the survival of whatever survives . . . . A strange *telos*: we are told simply, what survives survives. But this, it has repeatedly been objected, looks like a mere tautology. And at first sight, at least, it has also been repeatedly objected, a tautology seems to have no explanatory power, let alone the explanatory power that would be characteristic of a teleological account. For a teleological account distinguishes, and sets out as aimed at, a goal to which it can then relate the antecedent steps. Here, however, goal and steps are collapsed into an empty identity. (pp. 401-402).

Grene's point is that there seems to be a simple cause and effect relationship between variation in the environment and a variation in the gene pool of that population of moths. When by chance the trees become black, then the population becomes predominently black winged. After some time and concentrated pollution control (or other such chance events), the peppered moths predominate. There seems to be no "striving after" or "for the sake of" here. Young peppered winged moths can be understood to be under no different "imperative" in one environment than they are in another. The dynamics of this situation seem explainable in terms of the interrelationship between environment which, with reference to the system described, is controlled by chance. Perhaps situations such as this should be described solely in terms of chance and necessity. George Williams seems to support this as a method which may be employed.

> The mere fact of the effect's being beneficial from one or another point of view should not be taken as evidence of adaptation . . . Parsimony demands that an effect be called a function only when chance can be ruled out as a possible explanation.{22}

This seems fairly straightforward. One should rule out teleological (telenomic) {23} explanations when they do not capture the dynamics of a given situation. Even Aristotle agrees that the most one can do is attempt an explanation through teleology (P.A. 642a 16). But need this example actually cause Aristotle any problems? Grene generates her tautology from interpreting the end of a moth as "existing." Therefore, when a moth achieves its end, it exists (i.e., if it exists, it exists). But perhaps the telos is improperly described. If one were to say (consistent with Aristotle) that the end of moth larva is to become an adult moth, then if an individual moth does not reach this state (because it was eaten by a bird, for example), we say that its end was not actualized due to violence.{24} Indeed, whole populations can be described in this way without generating tautologies. In a blackened environment the peppered winged moths decrease in number (then do not reach maturity and/or reproduce) because of the intervention of another force (being eaten by the birds). Is not this explanation, in fact, what is actually being conveyed even by the short hand notation of equations, such as the Hardy-Weinburg formula?{25} The reason the number of peppered winged moths decreases is that great numbers of the peppered winged moth population at time t-one were unable to reproduce (their normal end) and so could not produce offspring to replenish the population at t-two.

This latter teleological account has greater explantory power than the non-teleological one, which asserted a simple relationship of increasing and decreasing numbers due to one variable, because this single function theory lacks the reason why the numbers of peppered moths decrease (in Larry Wright's terms it is non-etiological).{26} It merely states that in environment x, y population decreases and in environment z, y population increases. This explains the fact only. The teleological explanation accounts for both the fact and the reasoned fact, and consequently it is more complete (cf. P.A. 641a 12-15). This is not to say that an explanation is sufficient with only a teleological explanation, but only that any complete biological explanation requires a teleological account as part of the complete explanans.

The "reasoned fact" here is that which tells most completely why the peppered moths increase and decrease in number. A simple series of ratios does not capture this. Now some might argue that the efficient cause is ultimately the most important here because it describes the "triggering" factor which results in the decreases in population, viz., the change in environments. This kind of argument is reductionism which sees the efficient cause at one level and uses it to explain higher level phenomena. But this is surely not a complete explanation. The complete dioti is not known (is not understood as significant) unless the normal end of the system (including the normal ends of the constitutent parts) is understood. Though there may be a "trigger" which is critical to the system, the system as a whole, with its ends, must be comprehended before alterations become significant. The observer must understand the normal ends of moths in their environment before any of the other data can be comprehensible. Another group of examples which support this point come from the phenomena of co-evolution. Co-evolution works similarly to the moth experiment except that the variables are controlled by Nature. When two biological forms are dependent upon each other and one alters its somatic characteristics, it is sometimes the case that the other form may alter also in conformance with the first. An interesting example of this has recently been documented as occurring between a plant and an animal (the Heliconius butterflies and the Passiflora vines).{27} A further example of this might be seen in the evolution of animal immitators. These animals evolve to "immitate" the somatic coloring or structure of another animal, purely for defensive purposes or for advantages to be gained in the acquisition of food (e.g., the Monarch and the Viceroy butterflies).

It should also be mentioned here that this account only works when one accepts a multi-leveled scheme.{28} If one does not do this, problems occur when considering the status of a statistical population as an end to teleology. It might seem as if this statistical group were detached from the primary substance across an impossible schism. However, when one views the statistical population as merely a higher level of organization, this problem vanishes. The individual moth is to the population as the cells are to the moth. Indeed, the population may even be thought of as an individual (though for Aristotle this individual would have to be viewed as a unique collection).{29} Each level is defined by its particular functions which contribute to the degree of unity at that level. So it is that examples of population biology are found to be consistent with teleology.

Thus it has been shown that there are two senses in which teleology is operative in biological explanation: 1) In the way organisms actually develop (as a regulative principle), and 2) In the way that development and systems of development are to be understood.

In order to better illumine what this means in Aristotle's terms, let's look at the famous discussion in Phys. 2.8. The *aporiai* are clearly stated: 1) Why should Nature act purposively and not merely from necessity? 2) Why cannot the parts of animals come about from necessity? (198b 17, 24). Now in typical fashion these questions are addressed in reverse order. First, the teeth and other body parts come about regularly; all things which come about regularly are purposeful and not from chance; therefore, teeth and other body parts come about from purpose and not from chance.{30} Aristotle says that we do not use 'Chance' to refer to that which regularly occurs because regular occurance implies an habitual controlling mechanism, i.e., purpose. At this point one might ask why 'regular controlling mechanism' would be the same as 'purpose.' One might call purpose merely a divine intervention or an appearance useful only to the researcher to understand what is happening. But these two extremes, as has been shown at the beginning of the chapter, are not correct. It is clear that in this chapter of the *Physics* Aristotle dismisses both of these interpretations. First, divine intervention is contrasted with teleology at 198b 18, and at 199b 26 the supposed tie between purpose in Nature and an actual being which intervenes is dismissed: "It is absurd to think of the things which become that there is no purpose because one does not see a deliberative figure." The "decider" is not needed, therefore; why would a special case be made for *phusis*?{31} Such an assumption is "absurd." We need not have a supernatural agent for teleology. Likewise at 198b 29 the position of a mere 'as-if' teleology is linked with Empedocles whose position here Aristotle has shown elsewhere (e.g., P.A. 1) that he does not support.

> In cases in which all the parts come together just *as if* they were for a purpose, [some] such things survived--[they were] organized appropriately from spontaneity--those which were otherwise perished and continue to do so as Empedocles says his 'man faced oxen' did.

In Empedocles' position there is an 'as-if' teleology, that is, the event is not really teleological, but merely appears to be so. One might say that it was teleological purely as an explantory device. However, *in res* it is organized by chance. This cannot be, says Aristotle, five lines later in the passage (a position which runs through much of Aristotle and is embodied by the passage from the *Posterior Analytics*, "By Chance nothing comes to be for an end" APo. 95a 9).

Thus, these two interpretations: a) that teleology requires a supernatural, guiding force and b) that teleology is merely an heuristic device of understanding an event, both seem to be mistaken. The actual position is between these. Teleological explanation is the causal relation between Nature (and nature) and regular effects. Now the only plausible manner in which a modern can understand this is via physical and biological "laws." We say today that the behavior of matter is controlled by the laws of gravity. Biological generation is controlled by the principle of fitness. In a less quantifiable and more confused way Aristotle also thought that matter was controlled by its nature and that biological generation was controlled by a principle of the survival of the most fit (see discussion earlier in this section). The way that teleology works is outlined here by Aristotle, "Further, when an event has a completion, all the prior steps are for-the-sake-of that" (199a 9), and "[each] of the former [steps in an event] are for-the-sake-of the latter" (199a 15). Teleology is thus 'event' oriented (see appendix for a discussion of events and motion). The end of the event is said to be that to which the event strives. Teleology as a cause is therefore a cause which affects the event by affecting the end. If certain end products cannot live, then these are eliminated by the "survival of the most fit." By regulating the end product, Nature controls which organism will be (or remain) and which will not. Thus, in some event x, which has some individual or some group of individuals y, having property F at various temporal stages t1-t4; Fyt1 in x is regulated by any control upon Fyt4. If crooked tallons are requisite for some species of bird to survive, then this is a description of Fyt4, the end product of this type of bird's maturation. Such a relatively determinate end product will always or for the most part control the event to its completion (198b 35). An example from modern biology might be the interplay between chance and the evolutionary principle in mutation. One may justifiably say that the arrangement of codons of DNA might be random, but only certain arrangements will create an amino, acid and of these, only certain assortments will yield a protein. Certain aberations will yield nothing. Why? There is an established nested hierarchy which controls the end products of these microscopic events. These regulators act partly from the entrenchment of a successful organism/species (one which has a high fitness coefficient) and partly from structural constraints of physio-chemical laws. Any assortment is not possible and the parameters do not emerge by Chance. It is the determinate order of physical-chemical-biological "laws" or principles, which are at play, that fix (within bounds) what, for the most part, may be the outcome of a natural event.

It is true that Aristotle offers some bad arguments for teleology in this chapter, e.g.,

1. Intelligent action = Natural action.

2. Intelligent action is purposive.

3. Natural action is purposive (199a 10).

And

1. Art imitates Nature.

2. Art is for an end.

3. Nature is for an end (199a 17).

These arguments beg the question. However, when Aristotle talks about regularity in Nature he is on firmer ground. "For these and all natural [events] occur invariably or for the most part, and nothing [such as this] comes about from chance or spontaneity" (199a 1). And again when he describes the spider's web at 199a 26, Aristotle puts forth a theme he repeats in P.A. 1 on the regularity of end products implying a principle of regulation (cf. 199b 20ff.). Such a principle, by the above definition, would be teleological.

Now it is important to observe our "nature/Nature" distinction here. Hunks of matter may have numerous potentialities due to the amount of earth, air, fire and water found within. But since these may be directed anywhere, these physical natures do not exhibit any save the most general teleological descriptions. It is true that earth will not fall upwards in air since earth is heavy and its natural place is down. But besides such very general pronouncements, the nature of this hunk of earth does not determine whether it will be placed within my flower pot or be made into a brick. This is not to say that the nature of this hunk of earth is not important, for it governs what kind of brick can be made or what kind of plant may grow. What is important is that the individual and specific natures are seen in connection with their larger inter-specific principle, Nature. Such a principle might be the physical/biological constraints of an eco-system which might act to regulate the end products of its constituents. Now it is true that Aristotle does not often discuss such large interspecific principles, but he does use them as the organizational mode for _De partibus aniamlium_ in which common functions are treated together and not separately and repetitively such as man, ox and lion (639a 16). One could also appeal to passages which describe an overarching Nature (discussed earlier in this section). Aristotle recognizes the interdependence of species and the utility of various parts to the survival of the whole. This is presented in the context of a complex biosphere in which a single principle rules supreme: "Nature does nothing in vain" (P.A. 658a 9). This dictum passed through the ancient world as one of Aristotle's primary contributions to natural philosophy. Galen himself utilizes this maxim over and over to back-up his biomedical conclusions (cf. _Peri spermatos_ 1.5). Thus it is that Aristotle's teleology may be briefly characterized. I have termed it TE in order to highlight the relationship to conditional necessity (which will be termed ME). But even considered as itself, it stands as a useful truth which may be used to describe the biosphere.

Another contrast which should be illustrated is that between the theoretical sciences and the natural sciences (640a 1, 641b 10). From Metaph. 1025b 25 one knows that:

> Therefore, if all thought is either practical or productive or theoretical, the study of nature would be theoretical, but theoretical about things that have the power to move--and about being, according to the definition for the most part. It is inseparable from matter. Therefore we cannot fail to grasp how essence and the

> definition exist since without this our inquiry is rendered useless.

Phusike theoretikē deals with existent forms in matter. It concerns things that are in the world. These are tangible and have a connection to sense impressions. Thus Aristotle qualifies his statment by saying that the natural philosopher is interested in the logos as inseparable from the matter. If one were to study the logos alone, he would not be a natural philosopher. Though Metaph. K 7, 1064b also lists natural philosophy as a theoretical science, both passages make it clear that natural philosophy does not deal with abstractions (form separable from the matter), while mathematics does. A similar point is made in the De partibus animalium at 640a 1, though different language is used.

> The modes of demonstration and of necessity are different in the natural and the theoretical science.{32}

In this passage the theorētikon refers to mathematics or logic. The natural philosopher is seen in contrast to these as he deals with what is to be (esomenon, 640a 4) as opposed to what is (to on, 640a 4). The mathematician begins with "what is" (Phys. 200a 15-30). In mathematics one moves from things which exist such as lines and angles and so forth. From certain given premises we can deduce conclusions. These demonstrations are deductive and the necessity is absolute. In natural philosophy the case is different. One speculates on what is to be. If such and such materials are brought together, a house can be constructed. The mode here may be a deduction or an induction--either way the necessity is conditional. Since "what is to be" is a future event, there can be no absolute necessity. Thus, in natural science even if one is engaged in a deduction, it will be a different type of deduction than the mathematician uses because in the latter's case the conclusions will be necessary while in the former's they must be hypothetical.

One further difficulty in this passage is that Aristotle seems to be suggesting that the study of nature (biology here) deals with production of things (which will become). Seen against the passage from Metaph. E 1, it would seem that the study of biology is the study of a productive science. However, at 1025b 25 these appeared to be separate categories. In fact, natural philosophy was put into the category of the theoretical.

The proper way to read this passage requires us to remember the link between teleology to Nature. Nature is compared to an artisan. Aristotle is not saying that biology is not a theoretical endeavor, but merely that it is different from the solely theoretical activity of mathematics. As Aristotle says in Metaph. 996a 26, mathematics does not take ends into account. Mathematics is an artificial deductive system which concerns itself with following consistently from its postulates toward some absolutely necessary conclusion.{33} The conclusion of a demonstration (either mathematical or logical) follows from the premises necessarily. It demonstrates a truth which is atemporal and is divorced from action.{34} This is, of course, different from the biology. The biologist's activity can be theoretical in that he may be adducing first principles from observed systems through speculative activity. But the nature of that theoretical activity must bear the character of that which it studies. Since Nature can be compared to an artist or builder (productive activities), the biologist too, in his examinations, is engaged in an activity which bears the influence of its object, i.e., biology is teleological because Nature is structured that way. It is this connotation that Aristotle is implying in the passage P.A. 640a 1. The domain of biology is a productive-like activity whereas the domain of mathematics is a static, timeless realm of demonstrated truths. Therefore, biology is teleological and mathematics is not.

This theme is taken up again at 641b 11.

> Still then, from abstraction there can be no study of nature, since all things made by Nature are purposive.

This passage has bothered several commentators.{35} From the context of the last passage it is clear that Aristotle is again using <u>theoretikē</u> in a technical way, as he did in P.A. 640a 1, and <u>not in the</u> general way that he usually employs it as in Metaph. 1025b 25. Natural science cannot study that which is timeless, since its objects are all being constantly produced. Now one might ask about the essences or types derived by the biologist. In many ways these appear to be atemporal truths. Could it not be the case that these are the same as mathematical truths? Now it is true that once one discovers the essence of Man, he has a piece of knowledge which will always be true; this is because species do not alter their forms. This represents the broad theoretical aspect of biology noted in the last paragraph. However, within the technical usage that Aristotle is employing in this context (as he did also at 640a 1), the question is directed at the acquisition of knowledge and these objects which bring this about. In mathematics these objects themselves are timeless, and so the mathematician is concerned only with the process as it develops. The biologist, on the other hand, reflects the productive nature of his objects. Since "art is present in the objects produced by art" (640b 12), the person who studies such objects will reflect their nature in his method and results. Just as it was shown earlier that the end is actually in the organisms (e.g., the form being in the male seed, and the modern DNA examples), so also explanation of such a process was seen to be necessarily teleological. The *explanans* reflects the *explanandum*. Since the *explanandum* of biology is not essentially theoretical, but productive (*P.A.* 640a 3), its *explanans* will reflect this. This is why Aristotle says that *natural philosophy* is not concerned with abstractions. It is not the case that biology is not a theoretical activity, but it is a different kind of activity than mathematics since its objects, because of their natures, must be understood teleologically.

The understanding of teleology as a process in time leads Aristotle to the expression of causality within its realm. When a finite process begins, this does not mean that the sequence will continue until it reaches its end. If it did, then teleology could not explain the case of the peppered winged and black winged moths. The process may be interrupted at any moment; so that a beginning does not imply that its natural end will be reached. Aristotle's example is that a foundation does not necessitate that a house be built (G.C. 337b 32). From the beginning one cannot infer the end, though from the end one can infer a beginning. Causal connections in teleological explanations can only occur with reference to the *logos* of the process or to some process already completed. For *example*, if a UUU DNA configuration is transcribing a complimentary RNA chain, then the biologist can explain those interactions while they happen by referring to his knowledge of the logos of this process. This is equivalent to describing the event after it is finished. What the scientist may not do is predict with certainty what what will happen before it occurs, even if he knows how the process is supposed to proceed. For it is always possible that something will intervene and stop the process before it is completed.{36}

Each aspect of a teleological explantion reflects in some way the character of what it describes, viz., an emphasis upon the end or *logos*. Unlike necessity, which concentrates on mechanical interaction within matter, Nature, like Art, emphasizes the *logos*: *hē de technē logos tou ergou o aneu tēs* hulēs estin (P.A. 640a 32).{37} The logos or essence or form of a process is a regulative mode of our understanding. It also describes how some biological phenomenon actually is, "because the essence of man is what it is, therefore a man has such and such parts, since there cannot be a man without them"(640a 34). The essence or definition of Man causes (final and formal cause) a man to develop (via efficient and material causes) in a particular way. He must have certain parts, such as a head, heart and blood, and though he may lack others, such as arms and legs, the man does not function as well without them (640a 35). Their absence represents a deformity (a concept itself which presupposes normality, i.e., a fixed logos). Thus teleology, with its emphasis upon form as a unifying principle of any biological system, can be seen as an important component for biological explanation in general, both in Aristotle's day and in our own.

3.3 NECESSITY

Just what Aristotle means by events happening "ex anankē" is puzzling. This is because Aristotle writes about necessity in different contexts and the word itself has many uses. This section will concern necessity as it operates in nature/Nature. In the last decade some fine work has been done on this subject in an effort to begin to straighten out some of these difficulties. Such writers as Balme (1972), Kullmann (1974), Preus (1975), Sorabji (1980) and Cooper (1981,82) have offered interesting and varied interpretations of necessity and its natural applications.

Finding a proper place for natural necessity is extremely important for any exposition of Aristotle's biological science since necessity is one of the two primary modes of exposition in the biological investigations themselves. What tends to muddle or oversimplify discussions on natural necessity is the complexity of the concept. Several disparate strands of Aristotelian exposition are subsumed under this explanatory mode so that it is easy to blur these together or leave some out. My overall plan for this section is, therefore, to proceed through these several strands separately in an effort to tie them together into a single explanatory mode which will interface with TE (3.2) to offer a complete explanation in the biological sphere.

3.3.1 Absolute And Conditional Necessity

Aristotle begins his discussion of necessity at 639b 21.

> Necessity is not in all of Nature's things in the same way. Almost everyone tries to carry back their explanations to necessity, but they do not distinguish the meanings of necessity. There is absolute necessity, which is characteristic of eternal things, and there is conditional necessity, which has to do with all things that are generated as well as with artifacts such as this house etc.

He makes a more complete statement at 642a 2.

> There are then two causes: that-for-the-sake-of-which and the cause from necessity—for many things are generated from necessity. Perhaps one might ask which necessity is meant by those who say from necessity. For neither of the two modes defined in our philosophical treatises can be meant here. In things generated, however, there is a third kind. It is seen in things which come-to-be; as when nourishment is necessary, it is not according to either of the other modes, but we mean that without nourishment no animal can live. This is conditional necessity.

It is instructive to see these two discussions of necessity side by side since Aristotle rarely writes about necessity twice in precisely the same way. There are at least two types of necessities which are operable: absolute and conditional. These two differ in three ways: a) The objects over which each rules; b) The manner in which each operates; and c) The strength of the inference which each implies. These first two passages indicate the proper demarcation of (a). Absolute necessity's realm is the superlunar realm, and conditional necessity has regency over the sublunar. Thus the eternal and divine are ruled by absolute necessity (cf. Metaph. 1026b 27, G.A. 731b 25). The realm in which things are generated and pass away belongs to conditional necessity. Thus these two sorts of necessity can be separated by the scope of their respective domains.{38}

The second difference, (b), refers to the manner in which each operates.

> It must be considered whether necessity is absolute or hypothetical. Presently people place necessity just as if one supposed that walls are created due to necessity, because the heavy naturally descends while, the light ascends . . . without these a wall cannot come to be, but it is not because of these except in the sense that the material must be there, but it is for the sake of shelter and saving goods . . . . For example, why is the saw such as it is? In order that this can bring about such and such. This, however, would be impossible unless the saw were made of iron. It is thus necessarily made of iron if we are to have a saw and it is to perform this work. The necessity then is hypothetical . . . (Phys. 199b 33-200a 3; 200a 5-a 7; a 11-a 14).

Absolute necessity is connected to events which will occur without interruption. In this realm one can predict the outcome of an event before it has finished (see appendix discussion on events and motion). Thus, given a perspective at t1, one can know what will happen at tn [in principle]. This is because nothing can interfere with it. The elements are divine, and their motion is eternal and perfect (cf. G.C. 2.11). In contrast to this is be hypothetical necessity in which the conclusion is functionally related to the antecedent conditions. A saw in order to cut must be made of iron and must be sharp. The form is, "if x,y, and z, then g." The operational form, as P.A. 642 quoted above says, is that the nourishment is functionally necessary as an antecedent to life. A few lines later Aristotle uses the same kind of language he did in the Phys. passage. An axe has certain functional requirements which are necessary as antecedents given that a certain outcome has resulted. This is an important distinction between the way the two operate. Absolute necessity states from the beginning of the sequence and looks backward toward functionally necessary antecedent conditions (see discussions in Chapter Five). Once one has established a given end which is said to occur, then it must have been the case that such and such preceded it. Conditional necessity thus is not concerned with prediction.{39} An advocate of conditional necessity, therefore, would not define explanation in terms of prediction or in terms of a positivist model which subsumes current data into a set of "covering laws" to predict absolutely a given outcome. He would be more interested in showing why x, y, or z was functionally required to generate some effect. He would focus on the relationship of antecedent to consequent to determine which could be done without in order to get the bare bones skeleton of the event's structure. Explanation would thus take on a different meaning.

Thus, each examines a sequence from different directions. The distinction is made by Aristotle when discussing the construction of a house. Absolute necessity sees the genetic order as primary. The coming-to-be precedes the logos. The order is reversed in conditional necessity.

Distinction (c) refers to the strength of the inference that each implies from the vantage point of the beginning of the event.

> Because there are things which are invariable and of necessity . . . in the sense that they cannot be otherwise, and some which are not necessary always [absolute necessity], but are necessary for the most part, this then is the cause of the accidental. (Metaph. 1026b 27-32).

Absolute necessity is said to produce the force of not being otherwise. At t1, we can know with complete certainty what will happen at tn. However, the case is different with conditional necessity. Intervening conditions can arise and disturb the normal course of affairs. These intervening conditions, accidents, work against the normal, natural way of things and are thus contrary to nature (see 3.34). In biology, as in the entire natural realm, Aristotle is saying that, in principle, we cannot acquire certainty about the outcome of an event when viewing from t1. This may be due to the complexity of the biological realm or the indeterminateness of matter. But we look in vain for "laws" which will conform to the x => y variety.{40}

Thus we have shown three ways in which Aristotle has shown that necessity differs according to whether it is absolute or conditional. They apply to different domains and because they operate differently, the force which they exert is also different.

3.3.2 Necessity And Teleology

Now it was made clear in 3.1 that Aristotle intends to connect conditional necessity with TE to offer a comprehensive account of some event. However, such an enterprise would be impossible if necessity here were understood to be absolute necessity. Absolute necessity does not interface with teleological explanations--in fact there are some sharp conflicts involved. These difficulties have been highlighted by David Balme in his article, "Greek Science and Mechanism."{41} Balme shows three basic difficulties with such a relationship.

> It is illegitimate to draw a line between one group of purposeful activities (for the elementary properties can be nothing less) and another, and say that one works mechanically and the other teleologically. Secondly, even supposing that that were the case, how could Phusis be said to control or even to use interactions which were anyhow inevitable? The teleological explanation would be an illusion, for mechanism and teleology are mutually exclusive. Thirdly, Aristotle repeatedly insists that Anankē does not govern sequences. Yet what is mechanical cause-and- effect but a nexus of sequences in which the earlier event compels the later to happen? It is certain that Anankē cannot govern any such nexus, and the evidence which seems to point that way must mean something else. (p. 133).

These difficulties amount to: 1) Since mechanism [absolute necessity] and teleology are so different in nature, to impute the former to cover any subset of the latter would be arbitrary and unfounded; 2) If events are necessary, how can Nature "fashion" them for its purposes; 3) [haplos] Anankē is said not to govern sequences, yet this is precisely its domain, viz., the nexus of such sequences which constitute cause and effect.

These are powerful claims and are aimed at those who would want to posit absolute necessity in the natural realm and have it co-exist with teleology. A. Charlton, for example, in his notes to Aristotle's Physics makes just such a claim.{42}

> Although, however, the fourfold classification of causes is flexible in these ways, underneath it there seems to lie a firmer twofold division between two radically different types of explanation, one employing the concepts of matter, source of change and unconditional necessity and the other employing the concepts of form, end and conditional necessity.

Now, if we accept the criticisms which Balme offered above, the least one could claim here is that Aristotle has two modes of explanation which, though radically different and incompatible, are used to explain events. This notion would have to be based upon a dualism of mutually exclusive explanatory devices. One would choose the one which seemed to fit the best. But if the two are incompatible, then the scientist would be hard pressed to put together the various explanations in order to treat some novel case. Indeed, theory integration, a tool essential to scientific progress, would be impossible because one would not be able to fit together a teleological account and an account based upon absolute necessity to explain some third, novel case. At best, one would have two theoretical explanatory systems developing side by side. Both would be explaining the same things, but would be doing so from prospectives which cannot be integrated. This would seem to be a waste of time unless reality were actually so structured, viz., in a similar dualistic fashion--but this would violate Aristotle's doctrine of simplicity discussed in Chapter Two.

The dualistic model is at best awkward and an impediment to scientific progress because it blocks theory integration. At worst it requires us to make certain assumptions about a reality which is similarly structured (though we have no evidence to make such an assertion).

Much better would be to find a way in which one could have necessity and teleology in an integrated whole. This is done by adopting conditional necessity instead of absolute necessity as the operative necessity in _phusis_. Indeed, this tack has been taken by several modern commentators on Aristotle.{43} In order to see the comparative advantage hypothetical necessity here has over absolute necessity, let us turn back to the list of difficulties offered earlier by Balme. These cannot be solved using conditional necessity.{44} Let us address them in reverse order. First, the only necessity which might govern sequences [perhaps, see appendix]{45} is absolute necessity. But this necessity is limited to the superlunal realm. Therefore it does not concern us when debating about necessity in _phusis_. Indeed, by its very description, conditional necessity _must_ govern an event or sequence because it makes reference to antecedent and resultant conditions. These are trans-temporal and thus have to form a sequence. Thus this objection speaks against one imputing absolute necessity in Nature for only conditional necessity will satisfy this requirement.

The second point says that what is due to necessity cannot also be due to purposive creation. This again is correct when one thinks about absolute necessity. By its very name absolute necessity is seen to be a singular reason why something happens. It completely explains some effect. This effect is subsumed under a mode which gets one from cause to effect with nothing intervening. This is one way people understand the term 'mechanical.' There is a string of causes with effects each of which must follow. This sense of 'mechanical' has no place in Aristotle's natural science. It is confined to the superlunar realm. As I argue in Chapter Five, it is precisely this point which modern critics of teleology attack. They assert absolute necessity in natural events and properly point out that if there is an event p => q where p is the cause and q is the effect, then if q is necessarily [meaning absolute necessity], then q's existence is completely explained by p. Now this is a rather simplified model and does not take into account counterfactuals and the like (discussed in Chapter Five), but it does indicate the dimensions of the problem in a rough and ready fashion. If q is completely explained by p, then why bring in anything else? Nothing more is needed. To say that it is also for-the-sake-of-which in such a case is to speak of something superfluous: something unnecessary for q's coming-to-be.

There is great power in this objection. So much that I believe it alone properly excludes explanations of simple necessity and those of teleology from meaningfully being used jointly in order to provide an explanation. If one wanted to assert simple necessity in nature one would have two options: a) Declare that some cases are ruled by necessity and others by teleology, or b) Declare that the teleological account proffered was not in any genuine way causal, but merely an 'as-if' device used in order to clarify some seemingly purposeful situations. (The 'seemingly' here arises either from our mind's desire to seek purpose or in our language's hidden metaphors which always express such.)

Alternative (a) was discussed above and shown to lead to a dualism which has unfortunate consequences. (B) might be a way out, but it succeeds at the expense of weakening teleological explanation. This need not occur since we have conditional necessity at our disposal. Hypothetical necessity does not conflict with teleology since it is formed conditionally. If q is to have some properties or acts in a particular way, then p is a necessary antecedent. The very statement of conditional necessity involves teleology. Thus they do not conflict.

This, of course, speaks to the first objection as well. To say that necessity and teleology are incompatible can only be true if one misunderstands Aristotle's meaning for necessity or teleology in this context. If necessity means simple necessity and simple necessity moves infinitely in both directions, then teleology is inconsistent with such a necessity which does not admit a beginning nor an end.{45} Likewise, if teleology is seen as it was pictured earlier, as the future influencing the present, then there can be no sense of anything being necessary in the present except if one views that compulsion which comes from the future (certainly not a conditional necessity). But if one understands teleology as explained in the last section and necessity as hypothetical necessity, then the inconsistency vanishes. The very form of both modes is remarkably similar. Teleology defines its system through indicating the eidos. Then through metaphorical language (e.g., a purposive, animate Nature), it rationally describes how some events work "for the sake of" (i.e., to bring about) that end. Conditional necessity also accepts an end, but it explains the mechanical interactions which caused the end to be as it is. Both modes accept a limited system. Both view the system as a whole and try to explain how the end of some process was achieved through its antecedent events. Nothing could seem more compatible. One examines the purposive aspects from the metaphor of a creative Nature, while the other confines itself to finding the mechanical interactions between the components of a system and its environment.

It is not the case that one is identifying two different events and tagging one necessary and the other purposive. Rather, each event by itself can be seen from both perspectives. One event can be explained teleologically to determine why the event functions as it does, e.g., is it a part of some larger whole; does it provide a fundamental requirement for some organism's life? The same event can also be viewed through conditional necessity. In this way the biologist is explaining how the system or process operates. Thus, two fundamental questions of science, why and how, are answered through these modes of explanation.{46}

Aristotle's explanatory modes offer descriptive power. The emphasis is away from prediction because prediction is never a certainty. The dual use of these modes for explaining one event is illustrated by Aristotle as he analyzes the function of breathing.

> Exposition should be like this: for example breathing takes place on account of this [teleological explanation], while that takes place out of necessity. Necessity signifies that: (a) if that is for-the-sake-of-which, then (b) these must necessarily have that, and (c) that sometimes these are in this manner necessary owing to their

> nature. [For example], the hot necessarily goes out and comes in again when it meets resistance, and that the air should go in is already necessitated. And the hot substance within offers resistance as the cooling occurs and this causes the entrance of outside air and its exit. This is an example of the method about the things for which we must grasp the causes. (P.A. 642a 31-642b 4).

Aristotle states that respiration can be seen as caused by TE and by conditional necessity. He does not state the purpose here outside of indicating that there is a purpose (in P.A. 665a 15, Aristotle says the purpose of breathing is for the lungs and the lungs for the heart). One could say then that breathing has its purpose: to aid the heart. In addition to the teleological explanation one can also offer conditional necessity as an explanation of breathing. Sense (a) obviously indicates the 'conditional' part of the necessity. (B) and (c) are also aspects of conditional necessity. (B) indicates both its affinity to teleology in its form and its necessary nature; (c) defines how this necessity is brought about, viz., through the material nature. This material explanation offers the vehicle upon which conditional necessity is brought about. When one seeks to explain some event through purely material means (i.e., through the nature and motion of the matter), then one has a quasi-mechanical account. ['Quasi' because it is not associated with absolute necessity, as most mechanical accounts are, but is linked to conditional necessity.]

Conditional necessity requires that the system first be defined and then examined for what is necessary for the whole to operate. In this case, the whole is breathing: the movement of the air in and out of the lungs. The explanation revolves around the pressure that hot air creates. The heart area is hot, and so the hot substance within creates pressure when it comes into contact with the cold air (like a convection current). Thus, there is a continual inflow and outflow of air in the lungs.

The *heneka tou* accounts allows one to know *why* this respiration takes place, both for the lungs themselves and for respiration's relationship with the body as a whole. The explanation from conditional necessity (i.e., the material/mechanical explanation of a certain form) describes *how* this process actually takes place. Aristotle uses this method of dual explanation throughout his biology.

> It is clear that there are two modes of causation, and in our account it would be best to describe both if possible, or failing that, to try to make this clear for those who fail to state this tell nothing of value about nature. (P.A. 642a 14-17).

This theoretical statement is applied in De generatione animalium when Aristotle is describing female breasts and offers both kinds of explanations. As he summarizes one point he reminds his students of the two modes of description:

> In each kind of animal the place around the breasts is just such an empty space, and it so for both of the two possible reasons: it was formed such as it is (a) for the sake of the best, and (b) by necessity. (G.A. 776b 32-34).

Again at P.A. 646b 20-30:

> In the uniform parts, then, such dunameis are found apportioned out separately: one of the parts, for instance, will be soft, another hard, while one is fluid, another solid; one viscous, another brittle. In the non-uniform parts, on the other hand, these dunameis are found in combination, not singly. For example, the hand needs one dunamis for the action of compressing and another for that of grasping. Hence it is that the instrumental parts of the body are composed of bones, sinews, flesh, and the rest of them, and not the other way around.
>
> The cause which I have just stated as controlling the relation between them is, of course, a Final Cause; but when we go on to inquire in what sense it is necessary that they should be related as they are, it becomes clear that they must of necessity have been thus related to each other from the beginning. (Translation following very closely to Peck.)

Aristotle sees these two causes as compatible. Both are used to describe the same phenomena. However, there are other instances when he does not employ both (P.A. 677a 16), as one or the other serves his purpose (which is not always to offer a complete explanation of every part).

3.3.3 Material Explanations

In the last section there was some suggestion that conditional necessity needs to be incorporated with some of the causes in order to bring about a given effect. The rationale behind this is fairly simple. Conditional necessity is constructed out of two diverse concepts: 1) A conditional formulation, and 2) a necessary consequent. Now (1) is a mere formal structure, while (2) is an expression of a property which attaches itself to one of the terms. _Why_ does this property belong to the consequent? What is there _about_ the process which _makes_ it be the way it is? These questions call for a _description_ of some independent mechanism which can be explained apart from the _definiens_ in an effort to make it intelligible. This further link can be found in _Physics_ 2.9.

Here Aristotle distinguishes explanations of simple necessity from those of hypothetical necessity. As quoted in 3.31, the 199b 33 passage wants to claim that simple necessity, operating through the material cause, is not adequate for an explanation. What is wrong here is not that material causes are included for these are embraced in hypothetical necessity as well (200a 10), but that absolute necessity does not offer the adequate framework for the material cause. This framework or form is that which conditions the expression of the material cause.

> "The necessity then is hypothetical and not a result [as absolute necessity would dictate]. For necessity is in the matter and that-for-the-sake-of-which in the definition." (200a 14-15).

The pair of purpose with definition is merely the unity of form and purpose described in 3.2. This was termed TE. The other pair is necessity and matter. Here it is important to remember that we are speaking of hypothetical necessity. For matter has been described as aligning itself with absolute necessity by many commentators{47}. What is striking is that the phrase 'material necessity' is empty unless the type of necessity is also specified. Here it is clear that Aristotle is linking material causation with conditional necessity. It is my contention, following Weiss, Balme, Kullmann and Preus {48}, that the material cause is linked in _phusis_ to conditional necessity and not to absolute necessity.

But matter alone is not very helpful to conditional necessity. For matter without its movements would, strictly speaking, only include static nature as heavy, colored, of a certain shape and strength etc. A more complete account of matter will include its motions, i.e., the efficient cause. Aristotle frames the explanatory question at P.A. 640b 7, "What kind of thing is matter, and how is the universe formed out of it, and what caused movement . . . ." Here material and efficient causes are linked together. This combination under necessity creates the complete concept of hypothetical necessity. "It is clear then that we describe natural necessity as matter and its movements" (Phys. 200a 31). Natural necessity, which we now know as conditional necessity, is comprised of the material and the efficient causes which are grouped together just as we earlier had grouped the final and formal causes to comprise teleological explanation (TE).

Material/efficient explantion can account for something as 3.32 indicated through a unitary force which gives specification to the operation of conditional necessity. Thus, conditional necessity's functional character is explained with reference to some independent sources: the material and efficient causes. These are subsumed under hypothetical necessity to offer a powerful explantory device. They are quasi-mechanical in that the *explanandum* is accounted for without any reference to non-material, inscrutable forces. But it is a mechanism unlike mechanism today which depends upon absolute necessity. For Aristotle's necessity is hypothetical in character. I propose to term this complex mode ME, material explanation. It is clear from what has just been said that 'mechanical explanation' is also an appropriate appellation given the restrictions above.

As *Physics* 200a 14 indicated, these explanatory modes link accounts of matter to those of form. In this way Aristotle believes his explanatory framework is superior to his predecessors: especially Empedocles and Anaxagoras. In a much longer discussion in Metaph. 983b 6ff. Aristotle describes how many of the natural philosophers who preceded him posited a simple material cause to account for everything. Now in the *Metaphysics* passage there is criticism of these philosophers' theories as inconsistent with the very first principles they laid down. However, in this passage from *De partibus animalium*, Aristotle's point is that material causes alone are not sufficient for a complete explanation.

> It is not enough to say out of what it is made, e.g., as "out of fire," or "out of earth." If we were describing a bed or any other such thing, we should try to describe the form of it rather than the matter (bronze or wood)—or, at any rate, the matter, if described would be described as belonging to the composite whole. For example, "a bed" is a certain form in certain matter, or, alternatively, certain matter that has a certain form (a this in a this or in a this such); so we should have to speak of its shape and the manner of its form in our description of it—because the "formal" nature is of more fundamental importance than the "material" nature. (P.A. 640b 23-30).

The form is said to be more important than the material because it is the end of the material process. One might ask why each component is not equally important since the absence of either would mean that the thing could not be. This is true and Aristotle would not deny it. He shows his sensitivity to this issue in _De generatione animalium_ when he discusses the relative importance of the male and female roles in reproduction (the former is associated with _eidos_ through its own motions, and the latter is associated with _hulē_, G.A. 729b 5, 730a 25, 738b 25, 746b 10, 765a 1, 771b 15ff.). Though there is a sense in which the two, form and matter, are equal, this is only in a generative sense. In order for there to be generation of some x, both _eidos_ and _hulē_ must be present.{48} However, before generation there is a _logos_. That _logos_ is logically prior to the matter for it acts as a blueprint by which the developing matter follows. It is in this way that form is prior to matter (Metaph. 1029a 2, 1026a 5, 1070a 9; Phys. 193a 10; De An. 412b 10, 414a 14).{49}

Now _eidos_ should not be confused with _morphē_. The former represents the _psuchē_ of a thing, that which gives it "that what it was to be," or essence. This emphasis was missing, Aristotle claims, in such writers as Democritus who try to give morphological definitions of entities. Such "definitions" ignore the function and purpose of the part or organism in question. For example, a dead body is morphologically identical to a live one, but we do not call a corpse a man.{50} This is because the corpse lacks the essential qualities of a man which include the performance of certain functions. These functions are included in the complete _logos_ or _eidos_ of a man. An explanation of a bed which only tells 'from what' it was created, misses the most important aspects of description (640b 25). Material components have their place, but they only become relevant (as I have shown) after the thing in question has been described with reference to its purpose. So it is that one can expect a discussion of eidos (and the corresponding TE) before a discussion of hulē (and the corresponding ME).{51}

This was what was wrong with the phusiologoi.

> The physiologers talk like this when they describe the formation and the causes of the shape of animal bodies. Suppose we ask the carver "by what power was this hand fashioned?" Perhaps his answer will be "by my axe" or "by my drill," just as if we ask the physiologer "by what agency was this body fashioned?" he will say "by air" and "by earth"; except that the craftsman's account will be the better because he will not feel it sufficient to say merely that a cavity was created here, or a level surface there, by a blow from his tool. He will state why and the purpose for-the-sake-of-which he made the strokes and that will be for the purpose of the wood finally being formed into this or that shape. (P.A. 641a 7-15).

Early natural philosophers neglected to account for purpose and form. It cannot be the case that they did not recognize these aspects for they are preconditions of any other understanding (3.2). It is rather that they ignored these fundamental aspects of scientific method when constructing their own systems and writing their own methodological statements. They did not consider form or function, which are the components of the Aristotelian psuche. Aristotle repeats an argument from the De anima (414a 29) that it is only necessary to study those parts of the soul relevant to the parts or system under analysis (P.A. 641b 10).{52}

Aristotle sems to employ several applications to related aspects of logos and heneka tou. All of these names refer to some aspect of the theory in question, e.g., its essence which is first in logical order, though it is last in the order of formation (646a 25). That is, one must have blueprints in order to begin constructing a house. These blueprints represent the logos of the potential house. The house itself does not yet exist, but there is an idea of it by which the house may be constructed. The logos or eidos of that house is a completed whole qua formula. One can say, therefore, that before any construction of the house takes place there has to be such an end already existing. This is what is meant by the end existing first in real, essential order. (A biological example would be DNA in cell development, see 3.2.)

Once this end is in mind, the actual construction begins. To an observer watching the progress of the house, the beginning might seem to be the moment of ground breaking and foundation digging. This order is the sequence of generation. Generation proceeds until (if there are no interventions) the house is completed. To the observer it is at this time that the entire house "is." In one way this is true. It is not until the house is actualized that it is a house. But from another perspective, the house's logos comes first and the building next. From the generative perspective the process of formation precedes the end of the project because the end (seen from the perspective of conditional necessity) could not be were it not for the material generation which necessarily preceded it.

Thus we have accounts of both forms of explanation, TE and ME. These fit together to form a complete natural account of some event. Now there are some who have criticized models which attempt to reserve the natural realm for hypothetical necessity. One of the strongest attacks in recent years has come from Richard Sorabji (1980). Basically, Sorabji's arguments against seeing conditional necessity as the only form of natural necessity amount to three claims: 1. Hypothetical necessity violates Aristotle's program in the Posterior Analytics; 2. The contrast between necessity and teleology requires necessity to be absolute and not hypothetical; and 3. There actually seem to be cases in which Aristotle cites a natural necessity which is absolute.

Let us examine these objections. First, Sorabji claims that hypothetical necessity violates Aristotle's program in APo., "For there he thinks that scientists should prove with regard to various affirmative predicates that they belong everlastingly and necessarily to various natural kinds"(Sorabji, 1980, p. 145). The problem is, of course, that the APo. 1 (and part of 2) presents a model for a program which was not instituted for natural philosophy.{54} As discussed in Chapter One, APo. 1 offers a way to express a finished science in a clear, axiomatized fashion. What Euclid later did for geometry, Aristotle wanted done for all science. However, since knowledge never reached this state the project was never carried out. But could it be carried out in principle if biology depends upon hypothetical necessity? Where would the kath auto and kata pantos propositions fit in?

This is a more difficult point. I have argued that there can be no pure "laws" in biology for Aristotle, under the epi to polu dictum. If this is so, then any categorical statement about the working of the biological realm would be tempered by this condition and would be hypothetical. Such a position makes prediction only a percentaged estimate since there is always a degree of uncertainty. However, such a dictum as this can be stated itself with absolute certainty. Indeed, all varieties of meta-statements about the zoological world, e.g., survival of the fittest, the definitions of the genera and species etc., can be made with complete certainty even though the actual realm itself possesses a certain degree of indeterminateness. In actual biological events there is no complete certainty, but nevertheless, one can make a number of determinate judgments about this realm. Some of these which Aristotle makes are: A) In the natural realm things occur epi to polu; B) The Nature/nature dictums; and C) The classification of animals in which both genera and species are attributed determinate, affirmative attributes.{55} In each of these three types of endeavors Aristotle is able to maintain both conditional necessity and determinate categorical statements which are capable of being used in a program such as is described in APo. The key here is the difference in levels of discourse. When one gets away from individual swans and discusses 'swan' then one can, of course, operate with the full degree of logical rigor described in APo. This would involve the application of imperfect and perfect induction discussed in Chapter One. The degree of accuracy that one would have in the deductions made about groups would depend upon how well the first task was performed. Thus Aristotle's conditional necessity does not prevent him, in principle from making an axiomatic account of biology if and when enough generalized propositions about zoology can be formulated.

The second objection deals with the contrast between teleology and necessity. If conditional necessity is the only necessity operative in biology, then Sorabji claims that the two interface too well, i.e., that Aristotle implies there is a contrast between necessity and teleology, and with conditional necessity there is no real contrast, "It looks as if it is this other kind of necessity [absolute] to which Aristotle is referring or else he would not contrast necessity with serving an end . . . the necessity that is mentioned is contrasted with serving an end, and is pretty clearly not the relation of being hypothetically for an end" (Sorabji, 1980, p. 150).

These quotations indicate an interesting argument. Because Aristotle contrasts necessity with teleology, the two must be different. Conditional necessity and teleology are the same, therefore; conditional necessity cannot be the teleology to which Aristotle refers when he says that something happens "from necessity." Two important items must be noted about this argument. First, the word 'contrast' here is very misleading. Aristotle nowhere states that what he is offering in a 'necessary explanation' is an explanation which is entirely foreign to teleological explanation (as simple necessity would be). The only force of there being a "contrast" is that Aristotle offers one account and then the other. They are different, but the word 'contrast' suggests that these two are radically separate in the way absolute necessity and teleology would be. Lacking any such explicit statement about this on Aristotle's part it must be assumed that at the very least this word may be loaded in a way Aristotle did not intend. Second, it is incorrect that conditional necessity is exactly the same as teleology. There are striking formal similarities, but as has been shown in this section, conditional necessity also subsumes the material and efficient causes and so offers a striking contrast to TE, which only includes the final and formal causes. Indeed, this is what is meant by the full material explanation, ME. It is not at all redundant to offer an ME as I have characterized it along with the TE.{57}

The last point concerns a group of passages Sorabji feels must pertain to absolute necessity because they are either everlasting or seem to happen without exception. The former group of passages include various cyclic passages. There are the cycles of birth and death, and cycles of showers (G.C. 2.11, APo. 95b 37). These fall under the same category as the meta-statements above. They are not about any particular shower or any particular birth or death which would be governed by conditional necessity. Thus these passages are not problematic to ME. The latter group includes events which happen without exception. These include: thunder (APo. 93b 7-14), sap and leaves (98b 36-38), the movement of a limb (M.A. 702b 24), necessity in human action (N.E. 1147a 27, 30), necessity of plot development (Poet. 7) and animal desire necessitating action (Metaph. 1048a 14).{58} The last three of these examples do not really deal with natural necessity at all. Ethical concerns and dramatic artifice are not in the zoologist's purview. The examples of thunder, and the sap and leaves point to logical necessity in a demonstration and in definition respectively.{59} Of all the examples the De motu example is the closest to being simple necessity, but this is only because the movement comes about due to the material of the parts extending and contracting. If one believed that all material necessity was simple necessity, then this would be too. But I have tried to show above that there are many references to the material cause being linked to conditional necessity. This being so, there is no reason not to take this as another example of the same.

3.3.4 Kata Phusin And Para Phusin

A number of various meanings for necessity have been bandied about throughout this section. Most of these are encompassed in *Metaphysics* Delta Five where Aristotle defines necessity.

> "Necessary" means: a) That without which as a condition, life is impossible, e.g., breathing and nourishment are necessary for an animal, because it cannot exist without them. b) The conditions without which good cannot be or come to be . . . e.g., drinking medicine is necesary to escape from ill health . . . c) The compulsory and compulsion, i.e., that which impedes and prevents contrary to impulse and purpose . . . d) Again, what cannot be otherwise we say is necessarily so. It is from this sense of "necessary" that all others are somehow derived; for the term "compulsory" is used of something which is according to impulse . . . e) Again, demonstration is a necessary thing, because a thing cannot be otherwise if a demonstration has been absolute. (Metaph. Delta 5; following Trendennick).

Meaning (d) as stated seems to point to a common linguistic base for all the meanings of 'necessity.' This passage may well refer to simple necessity, though beyond the general linguistic characterization, nothing very specific can be said about it here.

The next type of necessity is that necessity which follows from a scientific demonstration (or any valid syllogism), necessity (e). This is the logical necessity of the conclusion following necessarily from the premises. Much of the *Analytics* is spent developing this type of necessity.

Meanings (a) and (b) are very close to the meanings of conditional necessity mentioned in *De partibus animalium*. In order to live, man must eat. In order *for an axe to cut*, its blade must be sharp. In order to perform some *ergon*, certain characteristics (*hexes*) must be present. This is because the function is *defined* with respect to its characteristics. The work cannot be performed without those characteristics. They are therefore necessary to said function. These are the connections one looks for in biological systems. When one observes mitosis, one is observing some activity of an organized system. The next thing the biologist wants to know is how it works, i.e., what characteristics or parts are necessary for it to take place. Through continued observation and photography one derives the idea of stages of mitosis. Then through staining and selective interference more of the necessary components are discovered. Aristotle adopts this usage of necessity when he discusses eunuchs in De generatione animalium. If one essential characteristic (arches) of the man is eliminated, the individual in question cannot perform the function that he was meant to perform. He takes on another role, that of a eunuch whose function differs in many respects from that of a normal man (G.A. 766a 29).

However, just because certain characteristics are necessary for some action to occur, it is not the case that the presence of these traits means that the function will take place. Meanings (a) and (b) speak of necessary and not sufficient conditions for F(x). For as mentioned in 3.2, there may always be intervening conditions which prevent F(x). This intervention is another type of necessity mentioned in Delta (c). Engaged in this way, necessity is working para phusin (against or contrary to Nature/nature). In APo. 95a Aristotle illustrates this by means of a stone whose natural tendency is toward earth. A stone unsupported in the air, kata phusin, will fall down. However, if the stone is borne upward, then it is moving contrary to its natural inclination. It is moing upward by force or violence. The stone has no choice; it must move para phusin. Therefore, when something interferes with phusis, necessity (c) dictates that the natural order be altered.

Besides force or violence, the indeterminacy of matter can cause an event to be para phusin and therefore in accord with meaning (c). In G.A. 778a 4 Aristotle says:

> Nature wants on the one hand to measure the generations and endings of things by their measurements. On the other hand, this is not exact due to the indeterminateness (aoristian) of matter and because of the many generative principles which are contrary to the natural process of generation and corruption which are so often the causes of things occurring contrary to nature/Nature.

Likewise at G.C. 336b 20:

> It often happens, however, that things pass away in too short a time owing to the comingling of things with one another; for, their matter being irregular (anomalou) and not everywhere the same, their coming to be must also be irregular, sometimes too quick and sometimes too slow.

In both of these passages the natural order which one might expect to occur is altered. The natural world is not determinate (as mentioned throughout this chapter). Necessity can step in and alter the normal sequence of events. By admitting this, Aristotle is asserting some kind of non-teleological cause for an event. It is not the case that causes always operate together. When they do the event is kata phusin. When they do not, it is para phusin. In these cases mentioned, certain material and efficient causes are working against the final cause of Nature. Mutations are an example of this. When hule's indeterminateness gets control of the process, it produces something unlike the essence of that species (G.A. 777a, 786b 30, 769b 10).

Necessity (c) can arise from an exterior force or from within the matter itself and alter an expectant sequence. There can be no determinism for precisely this reason. Though post facto, such aberrations may be explained, but they cannot, in principle, be predicted. One might wonder what Aristotle might say if confronted with better data by which predictions might be made. Because of his logical position enumerated in De Int. 9, it seems clear that Aristotle might admit that genetic predictions (for example) may become increasingly accurate, but that they can never--in principle--be exact. Biological phenomena are too complicated for this. Modern genetics seems to support this position by emphasizing a multiplicity of variables which are too great in number and complexity to precisely predict outcomes under existing models of explanation.{60}

Thus we are presented with a theory of necessity in Nature/nature. When it is purposeful it may be explained via ME and TE. However, the natures in Nature are not determinate. Matter and its motions are unruly. It is for this reason that effects can come about which have no purpose. They are accidents and contrary to the regular order. Because they violate the design of the system, they are "contrary to nature/Nature."

Notes to Chapter Three

1 This threefold division of the sciences is made in Metaph. 1025b ff.

2 It is impossible to show this with complete certainty. Many of these passages might make sense with Nature understood as a diety. However, as I have suggested, there is not any substantial support for this position either, which one would have expected had this been Aristotle's view. Among those who have also interpreted Nature as a non-diety are: Anthony Preus, "Nature Uses . . . " *Apeiron* 3.2 (1969): 20-33; F.M. Cornford, *Plato's Cosmology* (London, 1937), p. 39 n.1; F. Solmsen, *Aristotle's System of the Physical World: A Comparison with his Predecessors* (Ithaca, N.Y., 1960), p. 443.

3 Preus (1975) engages in an extended discussion, pp. 221-248, on the linguistic combinations which occur in the "nature" passages. Of particular interest is his discussion of *katachresis* (p. 227ff.) and the personal uses of nature (pp. 238ff.). Preus builds a case through an examination of the language that indeed there can be an overarching Nature who acts, but that it is a mistake to call such a force a diety.

4 It is interesting that Aristotle thought that the "Nature does not act in vain" passages at 658a 8 and 271a 33 seem to take on greater specification in the (presumably later) P.A. and G.A. passages. This might indicate that Aristotle worked out the details of an initial thought of Nature not acting in vain.

5 David Balme in an unpublished paper, "Aristotle's Natural Teleology," presented at the Conference on Aristotle's Biology, Marquette University November 19-21, 1981 makes a similar point about nature being the principle of survival of the fit. But Balme restricts this to individual natures transmitting what they have to their offspring. Balme would demur at the suggestion that there are group dynamics occurring here for

Aristotle, e.g., species involvement. Thus for Balme, there is no nature with a capital 'N.'

6
Representatives of those holding these two positions might be roughly grouped as follows: 1) Aristotle's teleological explanations depend upon attributing desires to all natural objects: E. Zeller, tr.as Aristotle and the Earlier Peripatetics vol.1 (London, 1897), pp. 337-349; Max Hocutt, "Aristotle's Four Becauses" Philosophy 49 (1974): 385-399. 2) Aristotle's teleological explanations are 'as-if' accounts. W. Wieland, Die aristotelische Physik (Goettingen, 1962), p. 261; and Martha Nussbaum, Aristotle's De Motu Animalium (Princeton, 1978), pp. 59-99.

7
Thus, evolution is impossible for Aristotle.

8
Two interesting articles on this and associated topics by M. Dummett are: "Can an Effect Precede its Cause?" Proceedings of the Aristotelian Society supp. vol. 28 (1954): 27-44; "Bringing about the Past" Philosophical Review 73 (1964): 338-59.

9
This much discussed passage has two good bibliographies to be found in: Richard Gale ed. The Philosophy of Time: A Collection of Essays (London, 1968) and William Ockham: Predestination, God's Foreknowledge and Future Contingents tr. M.M. Adams and N. Kretzmann (N.Y., 1969). Some of the best discussion on this subject can be found in: Richard Sorabji, Necessity, Cause and Blame (Ithaca, N.Y., 1980), ch. 8; G.E.M. Anscombe, "Aristotle and the Sea Battle" Mind 65 (1956): 1-15; Jan Lukasiewicz, (tr. as chapters 1-3) of Polish Logic 1920-39, ed. Storrs McCall (Oxford, 1967); A. Prior, "Three-valued Logic and Future Contingents" Philosophical Quarterly 3 (1953): 317-26; Jaakko Hintikka, Time and Necessity (Oxford, 1973), ch. 8; J.L. Ackrill, Aristotle's Categories and De Interpretatione (Oxford, 1963). Further texts of interest: APr. 1.13; Metaph. E 2-3, K 8; Phys. 2.4-6; G.C. 2.10.

10
Sorabji (1980) disagrees with those who would want to assert a physiological basis for Aristotle's teleology through references to DNA (p. 169ff). I disagree with this as I believe that Aristotle's dunameis can be understood as elemental

potentials--cf. Allan Gotthelf, "Aristotle's Conception of Final Causality" Review of Metaphysics (1977): 244ff. I likewise believe that pneuma is a material/efficient cause as well so that there is no contradiction in the modern analogue.

11
From the beginning there are at least two clear senses of innate: a) innate means that y was produced by A more tha by any other cause, and b) innate means y(z) was caused by A(x), given that x is the only relevant internal element to A. Definition (a) indicated that A as a whole contributed to y as a cause (to some degree) in comparison with other wholes. Definition (b) means that something x internal to A (that originated in a pure way from itself as active, dominant subject only) caused y with properties z. Definition (b) is clearly the stronger claim, but for Aristotle's teleological claim to hold only the weaker claim (a) is actually necessary (though both will be discussed). For an account of a similar definition of biologically innate see: Daniel Lehrman's, "Semontic and Conceptual Issues in the Nature-Nuture Problem" Development and Evolution of Behavior (San Francisco, 1970), pp. 17-48.

12
A similar example is explored in greater depth by David Hull, Philosophy of Biological Science (Englewood Cliffs, N.J., 1974), p. 26.

13
F. Jacob and J. Monod, "Genetic Regulatory Mechanism in the Synthesis of Proteins," Journal of Molecular Biology 3 (1961): 318.

14
Edward O. Wilson and William H. Bossert, A Primer of Population Biology (Sunderland, Mass., 1971), p. 22.

15
It is roughly teleological in that the traits of the parents tend to be passed to the children, but it also should be recognized that many factors are at the same time working para phusin, so that the process is not as regular as the more general traits which insure that the species will continue (though, of course, neither is necessary). Aristotle's account of variation is directed at the more specific, individual traits.

16
There are, of course, many of these similarities, but as one moves to more and more general categories the similarities become fewer and fewer. One that is shared by both types of cells is Cytochrome C. This has been documented by R.E. Dickerson, "The Structure and History of an Ancient Protein" _Scientific American_ 226.4 (1972): 58-72; "Cytochrome C and the Evolution of Energy Metabolism" _Scientific American_ 242.3 (1980): 136-153.

17
Le Blond remarks that the Metaph. 1032b 6 indicates that "le processus mental qui commande toute production humaine: le medecin qui veut guerir un malade doit avant tout savoir ou il veut en venir, c'est-a-dire saisir le logos, la definition de la dante, et c'est cette definition qui commande toute le reste" (p. 134, n.18). One can further put the point that one's knowledge of some x requires that x be definite, i.e., be x and not something else. If this condition is not met, then we can have knowledge of nothing. To put the problem in such a way is to say that the x is apparent to the observer as a whole. As a whole, it has a beginning, a middle and an end. The beginning and middle can be said to lead up to the end and are thus understood in that relation. If this is true, then the _logos_ of the process or system is a necessary precondition to understanding any of the parts of the process.

18
This is the conceptual "frame" discussed by Norwood Russell Hanson in his _Patterns of Discovery_ (Cambridge, 1958).

19
For a further discussion of a physiological interpretation of _dunamis_, see: A.L. Peck's introduction to his translation of _The Parts of Animals_, pp. 30-32; and Preus (1975), pp. 18-20.

20
Moliere, among many others (especially in his era), poked fun at what he considered a bogus form of explanation, cf. _Le Medecin Malgre Lui_ 2.4, "Aristotle says . . . the stoppage of her tongue is caused by certain humors, which among scholars we call peccant humors: peccant, that is to say—peccant humors."

21
Marjorie Grene, "Aristotle and Modern Biology" _Journal of the History of Ideas_ 33 (1972): 395-424.

22
George Williams, _Adaption and Natural Selection_ (Princeton, N.J., 1966), p. 261.

23
This term is used by Williams and others to mean the explicit recognition of the functional organization of a system. Colin Pettendrigh, "Adaption, Natural Selection and Behavior" _Behavior and Evolution_, A. Roe and G.G. Simpson eds. (New Haven, Conn., 1958), uses the term to mean the connection between teleology and evolutionary theory.

24
See the discussion in Phys. 2.6 on "spontanety."

25
For an explanation of this formula and its applications see: Bossert and Wilson, pp. 36-38.

26
Larry Wright, _Teleological Explanations_ (Berkeley, California, 1976).

27
Lawrence E. Gilbert, "The Coevolution of a Butterfly and a Vine" _Scientific American_ 247.2 (1982): 110-122.

28
For a discussion of multi-leveled organizations see William Wimsatt, "Reductionism, Levels of Organization and the Mind-Body Problem" in _Brain and Mind: Philosophic and Scientific Strategies_ ed. G.G. Globus (N.Y., 1975); cf. note 34 in Chapter Five.

29
This view of the population as an individual was suggested to me originally by David Hull. This particular position has been argued by Hull in "Are Species Really Individuals?" _Systematic Zoology_ 25 (1976): 174-91, "A Matter of Individuality" _Philosophy of Science_ 45 (1978): 335-360; and by M.T. Ghiselin in "A Radical Solution to the Species Problem" _Systematic Zoology_ 23 (1974): 536-544; cf. the discussion articles between Hull and Arthur Caplan in _Philosophy of Science_ 48 (1981): 130-152.

30
This is the implied argument at 198b 35, cf. APo. 2.2.

31
Of course another way to understand this passage is to assert that Nature might be divine even though we do not see any such figure. As I said earlier, one cannot completely rule out such a reading from a purely textual declaration alone, but the body of evidence as a whole does not seem convincing on this point; see: note 2.

32
There is some dispute on the *te . . . kai* here. Duering (1961), p. 215, takes this to mean that demonstration and necessity differ both in the natural and the theoretical sciences. Here the contrast is between demonstration and necessity as they differ in natural science and as they differ in theoretical science. I agree with Balme and Ogle here in seeing the contrast between natural and theoretical sciences, and that given this, one sees that the distinction lies in the necessity and demonstration attributed to each, viz., hypothetical necessity of nature vrs. the absolute necessity of eternal objects (either stars, or lines and angles).

33
Phys. 200a 13; APr. 76b 4; D.C. 306a 16.

34
The act of doing a problem can be seen teleologically, but this misses Aristotle's point here.

35
Duering thinks that there are two sentences here because he can see no relationship between the first half of the sentence and the latter half. He says that the fact that "No abstraction can be studied by Nature," has nothing to do with "Nature makes everything for a purpose." Because of this he suggests a period between the two clauses following Michael's suggestion. The problem with Duering's reading is that it makes too great a distinction between *hē phusikē* and *ta noēta*. It is true that mathematics is different from natural science, but this is only a difference in the objects of each. The disciplines reflect their objects so that the distinction alluded to here is actually reducible to a difference in objects. Igemar Duering, *Aristotle' De partibus animalium, Critical and Literaty Commentaries* (Goteborg, 1943), pp. 91-92.

36
Aristotle's whole notion of causality, as has already been mentioned is quite different from ours. All or most of Aristotle's examples seem to be about events which cover an extended period of time (see appendix). However, I would contend that there is a sense of material/quasi-mechanical interaction. This is seen in the definition of an event and in such specific examples as the "automatic puppets" see: G.A. 734b 10, 741b 5; M.A. 701b.

37
If one continues to the next sentence he will find, "And similarly with the products of chance: they are formed by the same process that art would employ." This sentence seems singularly out of place. This is precisely the difference between Art and Chance (_tuchē_): the former _is_ the _logos_ without the matter, but the _latter_ is without a _logos_ (_relevant_ to a design or purpose). The effects of chance _are accidental_, which is why it is a chance event (cf. Sorabji's discussion in part 1). Aristotle seems to be abandoning his distinctions he made in Phys. 2.4-8, see: Ross, Metaph. 2, p. 182. Michael paraphrases this passage, "All seems as if it were produced as it is in art." Duering calls it a footnote added later (1943, p. 85). This seems as reasonable a suggestion as any put forth for this very obscure passage (cf. APo. 95a 6).

38
Sorabji (1980) cites several examples of what he takes to be events which are ruled by absolute necessity in nature, e.g., cycles of birth and death and evaporation and showers and so forth. These will be discussed at the end of 3.33.

39
Many others have also attributed such an indeterminate position to Aristotle. For example Philoponus on G.C. 337b 22, Simplicius on Phys. 198b 5-6, and Michael on P.A. 640a 4. See also Preus (1975), ch.4; Kullmann (1974), pp. 292-93; Wieland (1970), pp. 264-66 and Balme (1972), pp. 80, 82.

40
Some writers, such as S. Sambursky, _Physics of the Stoics_ (London, 1959), pp. 71-73; I. Duering (1961), p. 215, have thought that this result would impede the progress of science. With respect to biology I will contend in Chapter Five that it does not. In fact, even modern biology can claim nothing more.

41
D.M. Balme, "Greek Science and Mechanism," Classical Quarterly 25 (1941): 129–138. Analogous arguments are often made. See: Jonathan Barnes, Aristotle's Posterior Analytics (Oxford, 1975), p. 22.

42
W. Charlton, Aristotle's Physics, Books I & II (Oxford, 1970), p. 115. Others who posit absolute necessity in nature include Sorabji (1980), ch. 9 and John Cooper, "Hypothetical Necessity," presented at the Conference on Aristotle's Biology, Marquette University, 1981, p. 6.

43
This position has also been taken by Balme (1972), pp. 76–84; Kullmann (1974), p.333; and Preus (1975), ch. 4, among others.

44
Absolute or simple necessity is clearly the stronger of the two so that any event which is governed by absolute necessity can be stated formally in conditional terms. "At A we can absolutely predict C," is an example of absolute necessity. However, if we waited until C came about, we could say, "Given C, A & B must have occurred." Thus a statement of absolute necessity can be expressed formally as one of conditional necessity. But what may be lacking here is the functional component. If one merely rewrites a statement of absolute necessity so that it is in conditional form, what one may lack is the functional dependence of the consequent upon the antecedent. The only dependence of such a formulation would be a temporal one. And even in cases where this formulation is *also* one of functional dependence, one would wish to state the proposition in the strongest possible terms, viz., absolute necessity. Thus every description has a proper formulation as either hypothetically or absolutely necessary.

45
One may even question this argument as it applies to simple necessity being incompatible with teleology in the realm of the heavens. Clearly, Aristotle intended teleology to work even here, Metaph. 1027a 25. Is it inconsistent? Perhaps. At least one must posit a strange sort of teleology to "save" Aristotle here. "Desire" tends the outer ring toward the unmoved mover. But since it is in a circle, this *orexis* causes the ring to turn. Now this did not need to have a beginning. One could easily say that it has always been that way. And just

because it harkens after a purpose or an end does not mean that it ever has to reach such an end; the process can be infinite and perpetually in the same state of actualization.

46
These questions roughly correspond to the first two scientific questions identified in APo. 2.1: "Is S P?" and "Why is S P?" Alan Gewirth has made the suggestion to me that these two statements here bare an interesting relation to Galileo's statement that he seeks the how and not the why.

47
Cooper (1981), p.6; and Sorabji (1980), ch.9 join Ross, _Aristotle_ (London, 1923), p. 79; J.M. Le Blonde, _Aristote Philosophie de la Vie_ (Paris, 1945), p. 136, in linking material necessity to absolute necessity.

48
Balme, p.80; Kullmann, pp.292-93 op.cit.; Preus (1975), p. 185; and Helene Weiss, _Kausalitaet und Zufall in der Philosophie des Aristoteles_ (Basel, 1942), pp. 75-93. Examples of the other view which would always link material necessity with absolute necessity include those in note 47.

49
Cf. Boylan (1983) on _eidos_ and _hule_ in Aristotle's conception theory. See also Preus (1977), Maryanne Cline Horowitz, "Aristotle and Women" _Journal of the History of Biology_ 9.2 (1976): 186-213; Johannes Morsink, "Was Aristotle's Biology Sexist?" _Journal of the History of Biology_ 12.1 (1979): 83-112.

50
Cf. Le Blonde, p. 154, n. 56.

51
Cf. Functions of the eye, P.A. 641a 4; G.A. 735a 7, 734b 24, 766a 8; De An. 412b 14; Metaph. 1035b 24.

52
This is, in fact, just what Aristotle does: P.A. 642b 31, 662b 25, 664a 19, 674a 10, 679a 10; G.A. 738b, 743b, 776b 30, 731a 24, 741b 25 etc.

53
It might seem strange advice for the naturalist to be admonished to study the soul of the animals he observes. This, of course, only reflects the emphasis that Aristotle puts on

form and function of the biological systems under analysis. It is only through these modes that the objects of the naturalist become intelligible at all. The more precise one's knowledge becomes of some species as an abstract whole the better able is the scientist to classify and explain the form which is before him.

54
Those parts of APo. 2 which talk about biological discovery, discussed in Chapter One, obviously pose no difficulty, as they are perfectly amenable to individuals and make no requirements upon the type of necessity involved.

55
Themistius, in fact, makes just such a biological interpretation of kata pantos at APo. 73a 34 (a chapter quoted by Sorabji to support his point). Themistius says that we can understand the predication of properties to animals, such as whiteness to swans, as proper instances of kata pantos. CIAG, p. 10.

56
Sorabji seems to recognize a tack similar to the one I am taking, but he dismisses it saying that it depends upon making a distinction between 'necessary' and 'always.' This is not the case in any of the meta-statements I have listed. They are stated in a categorical manner because they deal with groups and do not depend upon individuals. Biological indeterminateness resides within individuals and their interactions not in meta-statements about them.

57
In fact none of the examples which Sorabji cites, p. 149, nts. 15 7 16, shows a necessity which is either contrasted in the way Sorabji wants or which require absolute necessity. The G.A. passages at 789a 12 show teeth shedding from necessity, but this is within a teleological exposition. Indeed, the linkage is made explicitly at 767b 9 (another passage cited by Sorabji). The 766a 14 passage on necessity is full teleological language in the exposition of necessity (on conception). The best passages for Sorabji's point are 778b 16 and P.A. 670a 30. In the G.A. passages two necessities are mentioned (the second being the necessity which brings about accidents). This is also the necessity of the P.A. passage. See the next section for a discussion of this accidental necessity which is "contrary to nature."

58 Sorabji (1980), p. 147.

59 Cf. Barnes' account which views these passages as making (or trying to make) logical points and not ones about natural science, pp. 210-211, 242.

60 This is similar to a point made by William Wimsatt (1975) who has stated that Brownian Motion is incapable of a lower level explanation in terms of Newtonian reactions between the particles involved. This is because the large number of variables makes such calculations (even if they were possible) tedious and incapable of capturing the real dynamics of the event in question.

CHAPTER 4

THE PRACTICE

4.1 PLAN OF THE CHAPTER

It is not the purpose of this chapter to exhaustively treat each and every one of the 550 forms dealt with by Aristotle. Rather, the object will be to understand the method actually used by Aristotle when practicing biology in light of his statements made in P.A. 1. Special attention will be given to questions about the two sets of terms described at the end of the last chapter.

Aristotle asks at the beginning of the _De partibus animalium_ whether we should treat common attributes or treat animals singly such as Man, Lion or Ox (639a 15-19, see: 2.2). Now common attributes were shown to include both natural groups and common parts and their functions such as the heart and its functions.{1} Thus one is directed to examine natural groups and the common parts that function in each. Both of these modes of examination are important to Aristotle, and consequently; part of this chapter will be devoted to each type of Aristotelian investigation in an effort to discuss those passages which are representative of his work in these fields. It will also be of primary importance to illumine the critical relationships between empiricism and rationalism, and mechanism and teleology as they occur in Aristotle's biological investigations.

The primary works to be examined will be _De partibus animalium/_, _Historia animalium_ and _De generatione animalium_. It is in these works where Aristotle's most complete statements upon biological method may be found. _De partibus animalium_ and _De generatione animalium_ are among Aristotle's most sophisticated biological treatises, while the _Historia animalium_ offers a wide range of observations unmatched in any other

Aristotelian work. From time to time the other biological works, De incessu animalium, De motu animalium, Parva naturalia and De anima, may be mentioned, but usually as supporting material for these other works.

The first four sections (4.2-4.5) deal with groups of animals, while the last two (4.6 and 4.7) will treat the uniform and non-uniform parts respectively. Some comments may be made in passing on Aristotle's work in the the context of the history of science, but the primary objective in this chapter is to evaluate the way Aristotle goes about constructing biological explanations for things. Aristotle's mode of operation is the primary focus of this chapter and has guided the selection of passages and the manner of exposition.

4.2 INSECTS AND BEES

At the beginning of Book Four of H.A. Aristotle divides the so-called bloodless genera.{2}

> We now proceed to discuss the bloodless animals. These animals are divided into several genera. One genus consists of so-called 'molluscs' (Cephalopods) . . . another genus is that of the malacostraca . . . a third genus is that of the ostracoderms or 'testaceans' . . . the fourth genus is that of insects.

It is not surprising that Aristotle classifies insects as bloodless. Insect blood is without hemoglobin so it appears as white or yellow in color and could easily pass for some other variety of body fluid. Indeed, the entire circulatory system for insects is what is known as an open circulatory system. This means that the blood fills a body cavity and bathes all the tissues freely. The heart or dorsal tube is shaped altogether differently from primate hearts. It is a pulsating organ, but this fact would not be evident to someone unless he could dissect a live insect (something Aristotle could not accomplish without an anesthetic to render the insect immobile).

At 531b 20 Aristotle begins a detailed description of the common morphology of insects.

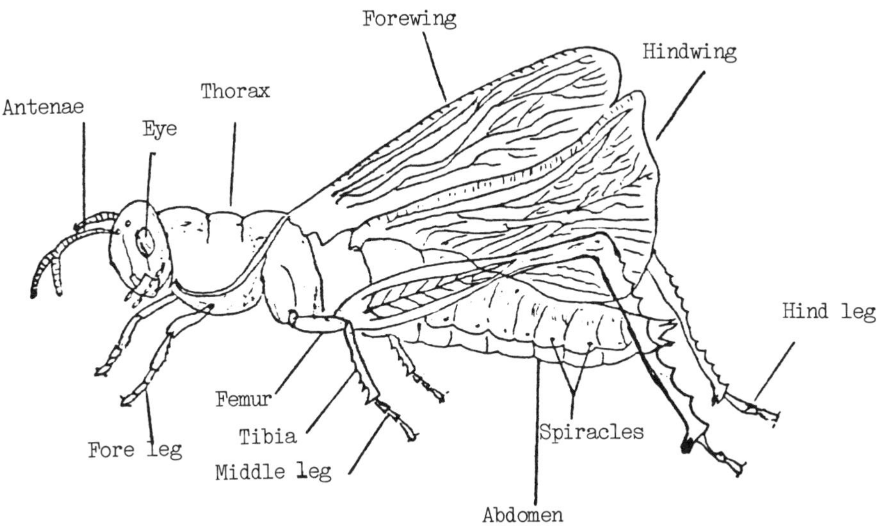

FIGURE THREE: Basic Insect Parts

> Insects have three parts common to them all; the head, the trunk containing the stomach, and a third part in between these two, corresponding to what in other creatures embraces chest and back. In the majority of insects this intermediate part is single; but in the long and multipedal insects it has practically the same number of segments as of nicks. (531b 26-29).

This description is very accurate and agrees with our modern classification (see figure 3).{3} The ability to distinguish the thorax from the abdomen indicates that Aristotle must have looked at insects in some detail himself.{4} The variation in segmented portions of the thorax shows this.{5} Another detailed observation is the variations in the digestive systems of insects.

> Internally, after the mouth, comes a gut. In the majority of cases it is straight and simple down to the outlet of the residuum. In a few cases the gut is coiled . . . some have a stomach also, and attached to this is the rest of the gut, either simple or convoluted as in the case of the acris or grasshopper. (H.A. 532b 5-6, 9-10).

The differentiation of the gut from the stomach indicates some dissection must have been done.{6} Aristotle was also interested in the position of the stinger in insects.

> In point of fact, such insects that have no stinger in the rear use this organ as a weapon, (and, by the way, such insects which are provided with this organ are not provided with teeth, with the exception of a few insects). The fly by a touch can draw blood with this organ, and the gnat can prick or sting with it. Certain insects are furnished with prickers or stings. Some insects have the stinger inside, as the bee and the wasp. Others have it outside, as the scorpion; and, by the way, this is the only insect furnished with a long tail. (532a 10-16).

It is interesting to note here in the _Historia animalium_, Aristotle (by and large) stays away from teleological explanations. His observations on insects in the passages noted are close to notes of recorded observations. However, in the _De partibus animalium_ when he talks about the stinger he is more concerned about offering a teleological account of its function:

> Some of the insects have a stinger as well, for defence against attackers. In some the stinger is in the front, by the tongue. In others it is behind at the tail-end. Consider the elephant's trunk: this is its organ of smell; but the elephant uses it as a means of exerting force as well as for the purposes of nutrition. Compare with this the sting of insects: when, as in some of them, it is ranged alongside the tongue, not only do they get their sensation of the food by means of it, but they also pick up the food with it and convey it to the mouth. (P.A. 682b 34-683a 5).

This passage is interesting in several respects. The sting is said to have one purpose when it is located in the rear: defense against attackers (one wonders why it could not have offensive uses too, though this possibility is not raised). The second function is to help in sensing the food and in some cases even in picking it up and conveying it to the mouth. It is here that Aristotle enunciates a principle of division of labor saying that:

> It is better, when it is possible, that one and the same organ should not be put to dissimilar uses; that is, there should be an organ of defence which is very sharp, and another organ to act as a tongue, which should be spongy and able to draw up nourishment. And thus, whenever it is possible to employ two organs for two pieces of work without their getting in each other's way, Nature provides and employs two. (P.A. 683a 21-26).

Here Aristotle notices an important principle, later to be used in evolutionary theory, that specialization of functions by a specific part or system allows that part or system to do a better job than an undifferentiated part or system. Aristotle's teleological point here is one of efficiency since Nature does nothing unnecessarily (that is, all parts are made to perform some special function; no part is created fortuitously, G.A. 744b 15). But it is also entirely consistent with our modern account of this important principle in biological organization.

Why does Aristotle assign these functions to the stingers of insects? It is probably the result of common experience. Bees, Wasps and Hornets seem to sting when bothered in some way and do not when left alone (the opinion of a bee keeper perhaps?). Aristotle's teleological conclusion about a rear sting for defense seems justified more or less by first or second-hand observation, but how does he determine that an anterior sting allows for sensation of food? Perhaps he came to this conclusion by observing that the stinger protrudes from the mouth. By analogy it would seem that it must be related to the tongue. But this is by no means conclusive. Why should there be taste or any other sense present in the stinger? Who is to say that the insect is capable of such a sensation? By analogy to mammals, perhaps, one can see where Aristotle got the idea that the mouth should be associated with taste, but what is there to prevent another part of the body from performing these functions? Aristotle incorrectly suggests that smell may be located in the middle of the body (P.A. 659b 15), so why cannot taste be located elsewhere as well?{7} Apparently such insects as mosquitoes, who convey blood to their body by means of an anterior sting, convinced Aristotle that the sting was instrumental in nutrition beyond mere conveyance. Though even if this is accepted it does not account for insects having taste, much less its location.{8}

As one might expect, a teleological explanation requires a mechanical one to compliment it. This Aristotle offers to explain why a bee's sting is internal while on smaller insects it is external.

> Sometimes the sting is well inside the body, as in bees and wasps. This is because they are winged, and a delicate sting on the outside of the body would be easily destroyed. (P.A. 683a 9-10).

The forces of wind and contact with flowers etc., would damage a fully external sting on a bee.

At H.A. 531b 30 Aristotle reports an experimental finding:

> All insects when cut in two continue to live, excepting such as are naturally cold by nature, or such as from their minute size chill rapidly; though, by the way, wasps notwithstanding their small size continue living after severance. In conjunction with the middle portion either the head or the stomach can live, but the head cannot live by itself. Insects that are long in shape and many footed can live for a long while after being cut in twain, and the severed portion can move in either direction, backwards or forwards; thus, the hinder portion if cut off, can crawl either in the direction of the section or in the direction of the tail as is observed in the <u>scolopendra</u>. (cf. 508b 4, G.A. 774b 30).

Whether Aristotle did this or not is impossible to know, but regeneration is certainly a widespread phenomena among insects. Not only do many insects have two extra legs that may be lost without impairment of their normal functions, but many times these legs will grow back (as well as damaged wings or antennae). Aristotle seems to have been more interested in the regeneration of one insect when it is cut into segments. The formula for successful regeneration given above is explained in more detail in P.A. 682a 2.{9}

> In insects the part where this control is placed, as was said in the first treatise, is situated between the head and the cavity where the stomach is. In the majority there is one such part, but in creatures like the Centipede, that is, which are long in the body, there are more than one: so if the creatures are cut up they go on living.

This passage shows that the mechanical reason for regeneration is that the <u>archē</u> of the insect, the stomach, is located in the middle section. Generally, of course, the <u>archē</u> is the heart. But in bloodless insects there is no heart. The analogue to the heart in these animals is the stomach. Thus the middle can survive with either the head or the feet (it also explains why the head and the end cannot survive alone). The <u>archē</u> of the bloodless animal is identified with the stomach. It thus becomes the crucial factor in regeneration.{10} In animals such as the centipede, which has multiple stomachs, there are more possibilities for regeneration since, if they are cut at a segment, both halves will remain alive. The mechanical principle dictates that on winged insects the middle plus any one other part (front or back) is sufficient for life and the regeneration of the mutilated member. In wingless insects, like the centipede, since there are several <u>archai</u>, several regenerations are possible.

This last possibility has some major ramifications for Aristotle's doctrine of <u>psuchē</u>. If a centipede can be divided and yield two centipedes, then which half has the soul, or do both have one-half a soul? Since both new centipedes will each acquire all the necessary centipede parts, it seems as if they have a complete soul (coming perhaps from separate stomachs?). Aristotle recognizes this problem and says:

> Now Nature's desire, [metaphorically speaking] is to make this part a unity in all creatures, and when she cannot, a plurality . . . the reason why finally they both

> coincide in one source and also begin from one source is this. The sensory Soul is, in all animals, one actually; therefore, the part which primarily contains this Soul is also one (one potentially as well as actually in the blooded animals, but in some of the bloodless animals it is only actually one [and potentially many]). (P.A. 682a 4, 667b 22-24, brackets indicate my interpolations).

A centipede has one actual soul, but potentially many since Nature has not been able to create a single unity from the centipede. This constraint will be one of the forces that works <u>para physin</u> (discussed in 3.3). That is to say the <u>indeterminacy</u> of matter affects the "purposive" creation of Nature.

At H.A. 550b 22 Aristotle goes through the generation of insects. Because of the size and habitat of insects this can be a very difficult problem for the entomologist. Other factors, such as parthenogenesis (where only a female exists and reproduces without a male) and polymorphism, complicate this problem immensely. Aristotle did recognize some of this complexity in <u>De generatione animalium</u>. Some insects produce young similar to themselves, while other produce only larvae.{11} Aristotle was quite aware of metamorphosis (H.A. 551a 13). But he did see generation as essentially a sexual interaction between male and female members of a species. Those insects that Aristotle attributes to non- copulation, the gnat and wool moth,{12} Aristotle believes were spontaneously generated. Some of these cases that he attributed to spontaneous generation are perhaps cases of parthenogenesis (though this phenomenon is recognized by Aristotle as it occurs in bees G.A. 760a). For other cases he says:

> So with animals, some generate from parent animals according to their kind, while others grow spontaneously and not from kindred stock. Of these instances of spontaneous generation some come from putrefying earth or vegetable matter as is the case with a number of insects . . . some out of dew falling on leaves, ordinarily in the springtime, but sometimes in winter when there has been a stretch of fair weather and southerly winds. Others grow in decaying mud or dung; others in timber, green or dry; some in the hair of animals; some in the flesh of animals; and some in excrements. (H.A. 539a22, 551a 1-5).

The rationale for such an explanation must surely be that Aristotle can see no other account. Insects seem to emerge from these unlikely places in which no insect has been seen for some time. It seems that they must have emerged spontaneously.

Obviously this is an argumentum ex ignorantia. Just because Aristotle cannot find a parent does not mean that there is not one. It simply indicates that Aristotle has not found it. Many aphids lay their eggs on plant leaves in late fall where they stay all winter coming out (seemingly from nowhere) in the spring.

Further, Aristotle describes insect copulation:

> Regarding Insects, some of them copulate. In those cases the young are generated from animals which are of the same name and nature as themselves, just as happens in the blooded creatures; instances of this are locusts, cicadas, spiders, wasps and ants. Others, although they copulate and generate, generate not creatures of the same kind as themselves but only larvae; and these insects moreover are not produced out of animals at all but out of putrefying fluids (in some cases, solids); instances of this are fleas, flies and cantharides. Others neither are produced out of animals nor do they copulate; such are gnats, mosquitoes and many similar kinds of insects. In most of the sorts which copulate the females are larger than the males; and the males do not seem to have any seminal passages. Speaking generally, the male does not insert any part into the female; but the female does so into the male upwards from below: this has been observed in many instances (and similarly as concerns mounting) the opposite in a few; but we have not yet enough observations to enable us to clarify them distinctly. (G.A. 721a 2-17).

On the surface this appears to be a passage supporting the value of biological observation. But whose observations? As mentioned, watching insects copulate in natural habitats is extremely difficult.{13} Compared to Aristotle's observations on insect anatomy they are markedly inaccurate. Most insects copulate in the same way other species do, and the male seminal passages in normal male insects are not blocked. Aristotle's tone is less definite in these lines. Perhaps this indicates that these observations are not his. Aristotle appeared more definite when he was recording the morphology of insects than he seems to be here. An explanation for this is that his former observations were his own and these are from others. Now every scientist must rely upon some set of accepted data, and Aristotle is not averse to this. But when the common opinion is not unanimous, then one must decide which is right and which is wrong. Evaluation of accepted observations is an important aspect of scientific methodology. Aristotle's language reinforces this interpretation. Since he refers to "most cases," it is clear that some cases do not conform to this account. Now this can mean "of all insects among species A-X, such and such behavior holds, and in Y and Z it does not," or more likely, "of all the reported observations I have, most of them describe this type of behavior, though others describe another." If this reading is correct, then Aristotle has made an evaluation of some observational accounts over others. It is true that he does not provide the reader (or his students, if these are lecture notes) with his criteria for evaluating the promising reports from the bogus ones. All that he says is that a definitive statement cannot be made as yet concerning this because more observational reports are needed to be conclusive. Whatever an adequate number constituted for Aristotle, it is clear that what he is constructing is some criteria for evaluating observational reports of others. It is true that none of this is formalized in writing, but there seems to be a rudimentary sense that many reports are needed to confirm a finding and that some are to be trusted more than others.

The conclusion which Aristotle makes here is tentative and subject to reversal upon further reported observations. One would expect that such shaky data would not be used to support any other points and certainly not be used to substantiate another theory (which he does in his exposition about pangenesis/epigenesis), since such substantiation is only as strong as its confirmed base. If theory x is shaky and is used to support theory y, then y's justification in this respect is equally as tenuous as x's (assuming no other independent justifications are offered). At any rate, it seems that any such use of x in a theory y should be noted by the theoretician so as not to mislead his audience into thinking x is a fact, when it is really a highly speculative hypothesis based upon insufficient observations. But Aristotle does just this at 723b 20 when he offers the tenuous account given at 721a 15 as an argument against pangenesis. He does not indicate the tentative nature of this statement, but on the contrary calls it his "weightiest proof of all." And though he does admit that it does not apply to all insects (l. 24), he does not indicate its tentative status which his earlier declarations indicate. In this latter passage Aristotle seems less concerned with presenting his material in a responsible way than he does with making his argument. (Aristotle in this context is not far removed from those thinkers discussed in Chapter One.)

Another instance of dubious methodology can be found in P.A. 682a 22, "All insects take but little nourishment, and this is not so much because they are small as because they are cold." The problem to be explained is why insects take so little food. Now it is interesting that this question does not prompt some discussion of 'little.' A beetle is smaller than a man or a dog and one wonders whether Aristotle thought that this might have something to do with its eating capacity. Physically it would be impossible for that beetle to consume at one meal as much as a man does. There is no place to put it. But perhaps Aristotle had some idea of proportionality in mind. He might be saying that a beetle in proportion to his size or weight eats less than a man in proportion to his weight. This reading is supported by Aristotle's observations of metamorphosis (H.A. 551a 15). While the caterpillar is in its cocoon it takes in no more food. In some species this stage can last a half year or more. If one estimates that feat in terms of food intake per body weight, it seems remarkable when one matches it against a non-hibernating mammal. Perhaps it is something like this that Aristotle had in mind as the phenomenona in question. His explantion for the insect's low food intake is that insects are cold. Now hot and cold are principles of digestion for Aristotle. Food is taken into the body and concocted (<u>pepsis</u>) by heat, thereby producing a change or transformation (<u>metabolē</u>) in the food to make it usable for the body (P.A. 650a 3, 670a 20, 25, 675a 10; cf. <u>pessein</u> as a rough synonym at 651a 20, 651b 25, 652a 5, 668b 5, 672a 1).{14} The critical element in this process is heat. Heat is generated from the heart and carried in the blood (653b 5, 696b 15, G.A. 772a). This heat controls the rate of concoction (G.A. 776a). One can think of the amount of heat in an organism as roughly equivalent to the modern concept of metabolic rate. When heat is high it will concoct a lot of food. When there is less heat, less food is required by the organism. An organism which is bloodless and heartless will be cold and thus require little heat, <u>a fortiori</u>, little food.

1. Hot and Cold are Contradictory Opposites--<u>a priori</u> premise.

2. Heat => High food intake--<u>a priori</u> premise.

3. Cold => Low food intake--from 1,2.

4. Blood + Heart => Heat--*a priori* premise.

5. -(Blood + Heart) => Cold-- from 1,4.

6. Insects = -(Blood + Heart)--observation.

7. Therefore: Insects = Cold--from 5,6.

8. Therefore: Insects = Low food intake--from 7,3.

This argument rests upon several loose connections. Premise one states the general truism of natural philosophy since Empedocles that hot and cold are opposite elements, such that if x is one, it is not the other. This is basically not problematical in this context. In premise two, Aristotle linked heat with high food intake. He does this by an analogy from cooking (G.A. 776a 2). Cooking transforms food. Some kind of transformation of food must take place in the body in order for it to be changed into nourishment. This process must be similar to cooking. (Perhaps Aristotle even observed that people grew hotter after eating a big meal.) All of these things made Aristotle want to posit the second premise. But it is clear that Aristotle has no empirically verifiable justification for this. Indeed it tends to the *a priori*. Though there are certainly empirical elements in the genesis of his heat/concoction theory, they are fragmentary and united by no direct observations. Analogy and indirect empirical evidence are very susceptible to misinterpretation.

It is this second premise which causes most of Aristotle's problems in the argument. The third premise follows naturally enough from the first two. The fourth premise might be obtained from observing the coldness of a corpse compared to the warmth of a living body. Since the major observable difference between the two is that in the corpse the heart has stopped pumping, it is perhaps understandable that Aristotle might believe the heart is the source of heat.{15} Premise five follows logically from four, and six could be the result of Aristotle's observation that insects have no red liquid which was in their bodies (a necessary property of blood for Aristotle).

One can find fault with Aristotle's method for arriving at premise two. It is contrary to his expressed empiricism. The explanation seems to have been generated more out of an over eager desire to posit an account than out of the more careful work shown on the other premises. This could be attributed to the difficulties involved with what he is trying to explain via dissection. But if the phenomenon is not phainei, then one is merely speculating about a biological problem. The point, of course, is not whether Aristotle "got it right" or not. Premises four and six are not "correct," but they seem to follow from an empirically oriented method in the way premise two does not.

What does one say about the entire argument? It would be distorting to characterize it as empirical or a priori. This is because it is a mixture of each. What is a more philosophically interesting point is to note how the various elements fit together and how they are justified.

Aristotle's observations on bees were almost surely acquired second-hand. His major discussion in the Historia animalium is told from a beekeepers's point of view with numerous references to how one should gather honey and drive the bees to their hives (623b 7-627b 22). In addition to this, some of the conclusions offered are distinctly un-Aristotelian in their nature. "Gaudy and showy bees, like gaudy and showy women are good-for-nothings" (627a 14). This sounds like an old saying popular among beekeepers. Aristotle reports, "Bees seem to take pleasure in listening to a rattling noise" (627a 15).{16} And then a few lines later Aristotle says, "It is uncertain, however, whether or not they can hear the noise at all, and also whether their [behavior] is due to pleasure or alarm" (627a 17).{17} Aristotle comments on the truisms of the beekeepers after first reporting them (cf. 554a 2, 623b 19, 31).

It is also interesting to note the lack of anatomical detail which Aristotle gives on bees compared to the more complete treatment afforded other insects. However, Aristotle is fascinated with the social structure of bees which is displayed in their reproductive behavior. In the De generatione animalium Aristotle begins 3, 10 with an admission of the difficulty of the problem, "The generation of bees is a great puzzle."{18} Aristotle runs through several possible explanations (perhaps those accounts offered to him by the bee keepers with whom he had conversed). Then Aristotle offers his own explanation (or one given which he believes to be correct). Aristotle presents the puzzling data. There are three kinds of bees: kings (what we now call queens), bees (workers) and drones. In a reproductive framework designed on a sexual model there should be only two kinds of bee: male and female. Aristotle reports that it is not even very certain which of these three kinds is male and which is female (759a 23). Now before any theories of reproduction can be produced, this most basic fact must be determined. Aristotle says that workers cannot be female because they have stingers and "Nature does not assign defensive weapons to any female creature" (759b 4). Nor can the converse be true (i.e., workers be male and drones be female), since workers take care of the larvae, and "no male creature makes a habit of taking trouble over his young" (759b 6). Thus, if the workers and drones can be neither male nor female, it must be the case that sexual copulation as it occurs in other animals must not occur in bees. Perhaps workers generate workers and drones, and kings produce kings (760a). Aristotle explores this possibility even to the point of coming up with a reason why female bees might be allowed a stinger (because they have male and female elements within them, 759b 30). (Aristotle again seems to be just trying to solve the problem in a logical way as he did with the insect argument in his discussion of pangenesis.) This account gives an explanation of the phenomena, however; as internally consistent as it might sound, it is rejected by Aristotle, "This however is denied by those whose business it is to look after these creatures" (760a4).

Those men who watch bees constantly deny Aristotle's _a priori_ explanation of bee reproduction and so Aristotle rejects _his_ theory on the basis of empirical evidence to the contrary (the reports of the beekeepers). Aristotle begins again creating an account consistent with his body of empirical data. The result (though a little obscurely phrased) has the king generating itself and workers, while workers only generate drones. The drones are incapable of producing anything (760a 33). This explanation shows Aristotle's grasp of polymorphism. It is obviously not the most comfortable explanation for Aristotle (his first one was more to his intellectual liking since it fit in better with the symmetry of Nature). But Aristotle does revise his theory to fit the observed facts. Even his last explanation, he declared, may be altered in the future if further observations show new answers.

> This then appears to be the state of affairs with regard to the generation of bees, so far as theory can take us, supplemented by what are thought to be the facts about their behaviour. But the facts have not been sufficiently ascertained; and if at any future time they are ascertained, then credence must be given to the direct evidence of the senses more than to theories— and to theories too provided that the results which they show agree with what is observed. (G.A. 760b 28-33).

Aristotle finishes this section on bees by trying to explain their social structure through his reproductive account. In reproduction the kings produce greater number of kings followed by workers. The drones produce nothing. Therefore, it makes sense that the kings should be the most powerful (760b 16). This argument by analogy is followed by another to show how similar links exist between reproductive behavior and dominance in Lions (760b 23). Aristotle also constructs two TE's to justify these accounts (760a 34, 760b 27). It is interesting to note that both the argument by analogy and the teleological account are given after an empirically based theory has been established. Aristotle seems quite willing to alter his teleological account of stingers in workers to accommodate the observed facts. The analogies are only offered after an acceptable theory has been presented. This is important, for as was shown in the first chapter among pre-Aristotelians, it was often the other way around. The analogies and rational accounts preceded the theory itself which was generated from the problem. These a priori hypotheses and analogies were what created ensuing internally consistent explanations. Aristotle has elevated observation to a point where it is possible to falsify one of his accounts. He is committed in this section to squaring his account with what is actually observed. It is why afterwards that the rational justifications have their place. This is consistent with Aristotle's theory in P.A. 1 as explained in Chapter 3. In that chapter a particular understanding of teleology proved to be the crucial element in unifying biological (and all scientific) investigations. One must understand the form of a system, its whole structure and function, before any account can be made. This "understanding," for Aristotle must be based upon observations of the systems or functions in question. Thus it is that the principles enunciated in APo. 1 and 2 are carried out in Aristotle's biology (for a more complete discussion of this see Chapter 5).

4.3 MARINE ANIMALS

Marine animals do not form a natural group. Aristotle sometimes treats marine animals together group by group for purposes of his exposition, but as he states in P.A. 1, classifying by habitat is not a good taxonomical principle (see: 2.3). Since there are so many different groups which are marine animals, this essay will focus on representatives of three bloodless groups: Cephalopods, Crustacea and Testacea; and two blooded groups: Fish and Cetecea.{19}

THE PRACTICE

Aristotle gives his most sustained treatment of the three bloodless marine animals at H.A. 523b 2:

> One genus consists of so-called 'molluscs'; by the term 'mollusc' we mean an animal that, being devoid of blood, has its flesh-like substance outside and any hard structure it may happen to have, inside. In this respect they resemble the red-blooded animals,—such as the genus of cuttlefish. Another genus is that of the malacostraca. These are animals that have their hard structure outside, and their soft or flesh-like substance inside, and the hard substance belonging to them has to be crushed rather than shattered; and to this genus belongs the crawfish and the crab. A third genus is that of the ostracoderms or 'testaceans.' These animals have their hard substance outside and their flesh-like substance within, and their hard substance can be shattered but not crushed; and to this genus belong the snail and the oyster.

Among the first group the reproductive behavior of the Argonauta (an octoped cuttlefish) has been praised as one of Aristotle's finest pieces of observation.{20} The sperm of the male comes in large spermataphores which the male transfers to the mantle cavity of the female with one of his arms. These spermataphores are held by several suckers at the end of one of the arms. This part of the arm detaches and is left in the mantle cavity of the female upon copulation. This process is called hectocatylization after the name 'Hectocotylus,' given by Cuvier, who mistakenly thought that this many suckered thing in the female cavity was a parasitic worm.

Aristotle gives three accounts of this phenomenon:

> The octopus uses his feelers either as feet or hands. With the two which stand over his mouth he draws in food, and the last of his feelers he employs in the act of copulation . . . some assert that the male has a kind of penis in one of his tentacles, the one in which are the (two) largest suckers. They further assert that the organ is tendonous in character, growing attached right up to the middle of the tentacle, and that the latter enables it to enter the nostril or funnel of the female. Now cuttlefish and clalmaries swim about closely intertwined, with mouths and tentacles facing one another and fitting closely together, and swim thus in opposite directions and they fit their so-called nostrils into one another, and the one sex swims backwards and the other frontwards during the

> operation. The female lays its spawn by the so-called 'blow-hole'; and, by the way, some declare that it is at this organ that the coition really takes place . . . In the case of the octopi, the male inserts his tentacle through the funnel of the female, and the fishermen allege that copulation is effected by means of this tentacle, but its purpose is really to link the two creatures together; it has no instrumental use so far as generation is concerned, because it is outside the passage [of the male] and outside his body. (H.A. 524a 4-5, 541b 9-15, cf. 544a 12; G.A. 720b 33).

In the first passage it is clear that Aristotle is affirming this process though the details are left unclear. The exactness of Aristotle's description of the parts of the cuttlefish indicates that Aristotle must have made some first-hand observations. Aristotle identifies five parts of the arm: 1. Arm (rachis), 2. Suckers (kotulēdones), 3. Sac of penis (ti epi tē rachei), 4. Bifurcated tip (akron dikroon), 5. 'Penis' (to oxu).{21} Aristotle's detailed observations here are remarkable.{22} The second passage from the *Historia animalium* offers more details of the actual process of copulation, but here one can note a certain hesitancy in his conclusion that the suckers are actually used in intercourse by his reference to other people's accounts. From the earlier passage it was clear that Aristotle has actually seen an Argonauta and probably one in the act of copulation. This makes his particular use of other people's observations different from his earlier invocation in the case of bee and insect reproduction. In that case Aristotle invoked other people's opinions (such as beekeepers) who had made many observations. As a biologist he was using their observations in lieu of his own. As a philosopher of biology, Aristotle was on good ground to employ an eye trained through many years of observation as an empirical source by which he could generate and verify his theories.{23} In this role Aristotle had to evaluate the raw data and pick out what generally was the case.{24} The other reports, if isolated, were relegated to the "spurious." However, in this case, Aristotle had his own and others' observations. There is an apparent conflict somewhere. It is not clear just where this conflict exists. It may be between two groups of observations by others, or (and I believe more likely) between Aristotle's observations and a nagging theoretical worry about how the arm becomes charged with spermatophores. If the arm is not directly connected with the seminal passage, then how can it be responsible for insemination? This is a difficult question. Aristotle thinks that he has observed such and such behavior.

Various fishermen have also confirmed his observation. But upon examining the Argonaut, Aristotle does not observe the seminal passage leading into the leg. One of two conclusions can be made from this: 1. The first observation was a misinterpretation of the observed facts, or 2. Somehow the spermataphores are transported to the end of the leg. Since no mechanism seems likely for the latter, Aristotle chooses the former.

Aristotle believes that every teleological process (which, as suggested in Chapter Three, is in every biological event, if not in every event in <u>phusis</u>) requires a mechanically determined process to bring about the end.{25} If the requisite mechanical means are absent, then the teleological description must also be wrong. In this instance it is rational deliberation about the mechanism of this system based upon observations which determine Aristotle's final judgment upon the matter. His evaluation of his dissection tells him that what he had earlier described from observation and reports could not be what was actually the case.{26} Perhaps the observed part of the tentacle left behind was a part of another proces, or indirectly related to mating, e.g., a courtship ritual. All of these possibilities could explain the observation without making it a mechanical part of the reproductive process. From Aristotle's dissection evidence, it became logically apparent that the arm could not be a part of reproduction (if spermatophores are necessary for reproduction and no seminal duct is present in the arm, then the arm cannot be directly involved in reproduction).{27}

Aristotle's final conclusion does not agree with our modern account which supports his earlier finding, but his method is sound. If the mechanism is not there, it suggests a misreading of the original data. Behavioral observations need to be supported by physiological mechanisms which can be explained in systems that act according to conditional necessity.{28} Aristotle's answer might be wrong here, but an application of his method will allow one to arrive at the correct answer.

Aristotle's dissections of Cephalopods is quite detailed (H.A. 524a 10- 525a 15). He accurately observes that the bone of the cuttlefish is not simple like other bones, but "spongy." This is a correct depiction of what is now known to be a complex structure.{29} Other descriptions, such as the duct from the oesophagus to the mantle cavity, are not accurate. Aristotle must have connected several similar independent passages.{30}

THE PRACTICE

Of the Crustaceans Aristotle gives accurate descriptions of several species that one might find in shallow pools alongside the shore. This indeed may have been Aristotle's method of examining the behavior of these animals. He notices certain habits of such creatures as the hermit crab's expropriation of empty shells for a home (H.A. 529b 20), and the sloughing of a lobster's shell (601a 10). Other observations include the bifurcated first foot of the female crawfish compared to the undivided foot of the male (hence a recognizable bimorphism, H.A. 526a).{31} Other observations, such as the method by which Crustaceans take in water (Aristotle claims that it is via an opening close by the mouth, 526b 18, 527b 17), are incorrect, but such an observation is difficult. It is in the _De partibus animalium_ where Aristotle makes his more important statements about Crustaceans. At 683b 25 he breaks them into:

> There are four main groups of Crustaceans, called: 1. Lobsters (_karaboi_), 2. Crayfish (_astakoi_), 3. Prawns and Shrimp (_karides_), and 4. Crabs (_karkinoi_). Each of these contains several species which differ not only in shape, but also in size, and that considerably, for some species are large, others extremely small.

Now what is interesting in this chapter is Aristotle's use of _a priori_ principles to explain the functions and sizes of various parts. The first example concerns the reason why lobsters have tails while crabs do not.

> The _carabi_ find a tail useful because they are swimmers: they propel themselves with it as though with oars. A tail would be useless to the _carcini_, which spend their lives near the land and creep into holes and crannies. (684a 3-7).

This sounds a little like the point Aristotle was trying to make in refutation of Anaxagoras' statement that Man is rational because he has hands (P.A. 687a 6). The part results from the function and not _vice versa_. The lobster has a tail because it is a swimmer. The crab, not being a swimmer, has no need for a tail. This sounds very neat, but upon further examination one wonders how the lobster became a swimmer. If it had no tail it would not be a swimmer. This case is a little different from the rationality/hand example because in that instance hands are not linked to rationality (a mutant race of Men without hands could be rational). But a species of lobsters that had no tails could not be swimmers. In this example the part seems not to be subservient to the function, as was the case in the rationalty/hand example. If one were to put function over the part, then one would have to say that lobsters acquired a tail in order to swim. But this seems strange since without a tail the lobsters would not be swimmers. How would the notion of their being swimmers come about? Is it that part of their esence is being "swimmers"? But what would it mean for a tailess species to have "swimmer" as its essence? Of course a species does not alter and evolve for Aristotle, but such a process can be legitimately employed in a thought experiment to determine priority. One would expect Aristotle to respond that part of the _logos_ of a lobster is as a swimmer and thus "swimmer" is logically first. Practically, before a lobster can swim it must grow a tail. The part is first in Nature (see: 3.2). It is here, most significantly, where modern evolutionary teleology (or telenomy) diverges from Aristotelian teleology. For a modern account requires one to accept certain functions as being essential attributes of some species. Each species has a form, and all growth must be understood with reference to that form (see: 2.5). The second example concerns the claw of lobsters. Aristotle says:

> These claws are not for the sake of lobsters, but serve instead of hands, for catching and holding. That is why they bend in an opposite direction to the feet, which bend and twist toward the concave side, while the claws bend toward the convex side. This makes the claws useful for taking hold of the food and conveying it to the mouth.

Aristotle notices that claws are used for eating, and by applying the dictum that Nature does not put one part to dissimilar uses (683a 24), he shows that it must be the case that claws are made for eating. Now one might ask whether the claws might also be made for walking, but Aristotle responds that since the claws bend in an opposite way from the legs, the claws must not be for walking (for only elephants are arranged in this way (_De motu_ 712a 12).

It is observed, however, that lobsters _do_ use their claws for locomotion which Aristotle must claim is _para phusin_.

> In lobsters only, whether male or female, it is a matter of chance which claw is bigger. They have claws because they belong to a group which has claws, and they have them in this irregular way because they themselves are deformed and use the claws not for their natural purpose but for locomotion. (684a 31-35).

Aristotle in this passage is linking two irregularities: the size of the claw (which is "deformed" because *kata phusin* the right should be bigger than the left, since the right side is superior to the left, P.A. 684a 25, *De motu* 706a 21), and the use of claws for locomotion. Since the lobster does use its claws for locomotion, when the claws are designed for eating only, the locomotive use of the claws is contrary to Nature. That the lobster is to Nature in this respect is no wonder, Aristotle seems to be saying, since it is contrary to Nature in other respects as well.

This assigning of a single function for a single part (i.e., each part having a distinct and separate *ergon*) by Aristotle is a dogmatic application of his general rule. This rule, it is interesting to note, was stated in the context of insect stingers that protruded from the mouth. This stinger was said to have a double purpose, i.e., to be an exception to the general rule which was then laid down. In that case, Aristotle was more flexible with his theoretical generalization. Observation counted for more than his general rule. In the cae of lobsters, however, Aristotle seems to have thrown away his respect for empirically based generalizations in favor of his theoretical ones. If all lobsters use their claws for locomotion, how can *that* be contrary to nature/Nature? One would expect events that are *para phusin* to be sporadic (see 3.3 and Chapter 5) and certainly not convertible properties of the species.{32} It seems that Aristotle is at a loss to explain why the lobster's right claw is not larger than the left as is the case with crabs. The right/left symmetry fits in with a general principle of Aristotle's, and so it is internally consistent with the general theoretical hypothesis that Aristotle has laid down. But to use those hypotheses and their conclusions in this instance, where the observed data do not confirm the theory, is an a *priori* move by Aristotle. He is putting data second and *explaining* away irregularities as simply being contrary to Nature. This is one place where Aristotle's empirical method is not carefully followed.

In the third case of *Testacea* Aristotle makes some interesting comments on sea urchins.{33}

> The sea urchins all have the same number of ova--an odd number, five, indentical with the number of teeth and stomachs which they have. This is accounted for by the "ovum" not being really an ovum (as I said before) but simply a result of good nourishment. The "ovum" is found in oysters too, though on one side of the body only. The

> same holds true with the sea urchin Now the sea urchin is spherical and is not just one flat disk like the oysters. Its spherical shape is the same all around; it is equiform in all directions; hence, of necessity its "ovum" is correspondingly arranged. Because this creature's perimeter is not, as in others, non-equiform: they all have their head in the center, whereas the sea urchin's is at the top. Yet even so the "ovum" cannot be continuous, since no other of the Testacea has it thus; it is always on one side of the disk only. Hence, since this is a common property of all species of Testacea, and the sea urchin is peculiar in having a spherical shape, the result follows of necessity that the sea urchins cannot have an even number of ova. If they were even, they would have to be arranged in diametrically opposite positions, because both sides would have to be alike . . . Therefore there must be three or five or some other odd number of ova in the sea urchin. If there were three, they would be too far apart; if more than five they would be quite continuous; the former would not subserve a good purpose, the latter is impossible. Therefore the sea urchin must of necessity have five ova. (680b 4-21, 25-28).

Again, Aristotle offers a rather odd argument. First, Aristotle does not believe that these ova are really ova, but masses of fat (see Ogle's note, 680a 13). They are actually the ovaries or testes. This is an observational error of minor importance here. Second, Aristotle believes that because the oysters and other Testacea have an asymmetrical distribution of their ova, the sea urchin must also. This argument amounts to saying that some class x has a membership requirement of having properties F and G. Now A, B and C are known to be members of x. A and B are known to have F and G, but in the case of C only F is known and it appears that C has another property G'. According to Aristotle, C ought to have G if it is to be a member of x. Sea urchins belong to the natural group of Testacea because of general functional and morphological considerations (F). But just because of this, need G (the distribution of ova) be attributed to C as well? Perhaps G is not a membership requirement of x. It could be that G is an accidental property of A and B and not an essential attribute of "x-ness." If this were the case, C need not have G. But if G is a valid membership requirement, then why not simply exclude C from x? Why say that the observed G' is an aberration? Why go into an ad hoc argument about how an exception might be made in this single instance (because a sea urchin is closer to being round while an oyster is closer to being flat)? Indeed, there seems to be no justified scientific reason for Aristotle's move here at all. Observation tells that the distribution of "so-called ova" are symmetrical in the sea urchin and asymmetrical in the oyster. What Aristotle wants to do is claim that "oddness" alone makes the number of ova asymmetrical in some way. An even number is the most symmetrical since half are on one side and half are on the other. An odd number defeats this possibility. An odd number makes bilateral symmetry impossible. Then Aristotle shifts again, thinking this answer satisfactory, to a discussion of what odd number ought to be used: three, five, seven . . . ? Aristotle decided that five is the best number because it distributes the ova "best" in the sea urchin. This last tack is imprecise. To say that three would make them "too far apart" is purely subjective opinion based on some personal aesthetics. What constitutes "too far"? Unless "too far" is expressed in relation to performing some function rendered mechanically impossible because some degree of proximity needs to be maintained, Aristotle's judgment here has no biological significance and is a hypothesis based on a random preference not pertinent to the performance of the system in question. But also, this last move is beside the point. It does not address the problem it has been given to support, viz., how the sea urchin, whose ova seem perfectly symmetrical in their distribution, are really asymmetrical because of "oddness" or some other property. The original comparison with oysters was meant to illustrate asymmetry as uneven distribution—not total numbers, be they odd or even. It is also apparent that the sea urchin having five stomachs or five teeth (680b 28) has nothing to do with Aristotle's original assertion.{34}

Aristotle has three logical alternatives when confronted with C's not having G: 1. Declare that G is not an essential attribute of x, 2. Declare C not to be a member of z, or 3. Declare that C really does have z in some strange way. Of all three alternatives (3) seems to be the most *a priori* and least in accord with a methodology which proclaims that sensory observations must be taken seriously and are capable of falsifying a theory. One would rather suppose that alternative (1) offers the best way out for Aristotle. If he believes a sea urchin should be included with Testacea, then it would seem like the best alternative would be to declare the asymmetry of ova to be an accidental property of many but not all Testacea. This move stays closer to the observational data and makes no *ad hoc* changes in theory (invoking no "special definitions" etc.).

Aristotle's descriptions of fish are as rich and detailed as his descriptions of the "non-blooded" marine animals. At H.A. 507a 2 Aristotle enters into a generally correct description of the internal organs of fish.{35} Then at 601b 8 Aristotle says that most fish enjoy the rain and hate the cold. This information seems primarily aimed at fish that live near the surface of the water. These fish are the ones sought by fishermen, who are undoubtedly the source here. Freshwater fish, who live near the surface, are ripe for catching during a rain storm because it is at this time that many of them feed. But these species only represent a small portion of the total number of fish. As for cold, each species of fish tends to be very sensitive to its temperature range and does move accordingly, but this is certainly not the proximate cause for fish migration as Aristotle seems to think at 597a 15. Also, Aristotle's claim that the number of days for fish gestation is always a multiple of seven (570a 30) is, of course, incorrect and an example of making a generalization from a too restricted sample space. In another case, however, Aristotle shows a willingness to discard hasty generalizations. This instance concerns the observation that fish of different species do not mate.

> Observations would lead to the general rule that no one variety of fish mates with another variety. The angel-fish, however, and the batus or skate appear to mate with one another. And there is a fish called the rhinobatus, with the head and front parts of the skate and the after parts of the rhine or angel-fish just as though it were made up of both fishes together. (H.A. 566a 26-30, cf. 540b 11, G.A. 746b 5).{36}

Generalization of many cases leads to the rule that among fish only like kinds mate. But having observed a fish with the head of the batus and the back parts of the rhine, Aristotle concludes that it is possible that this "rhinobatus" is a cross between two fishes. If this is the case (and Aristotle seems to think that it might be), then his general rule is incorrect as a universal generalization and holds, if at all, as a rule "for the most part."{37} Though Aristotle does not pursue the matter, it is clear that in this case observation governs the establishment of a generalization.

One conclusion of Aristotle's which is quite interesting is his argument that fish sleep.{38} He says:

> But it is from the following facts that we may more reasonably infer that fishes sleep. Very often it is possible to take a fish off its guard so that one can catch hold of it or to give it a blow unawares. All the time that you are preparing to catch or strike it, the fish is quite still but for a slight motion of the tail. It is quite obvious that the animal is sleeping, because if any disturbance is made during its repose, it moves just as you would expect in a creature suddenly awakened. (H.A. 537a 13-17).

One cannot infer that fish sleep from their eyelids being closed, as one does with other animals, since fish have no eyelids. Rather, one observes that fish are motionless and are more easily captured at one time than at another. Of these two pieces of evidence, the first is somewhat suspect since one cannot be sure that the motionless fish is not simply watching something or engaged in some other activity unknown to the human observer. The second piece of evidence is better since it indicates that the sensory powers are impaired or else they would not be captured so easily.

The way this argument works is through an analogy with other animals. Sleep, is, by definition a state in which an organism's sensory powers are not sharp. This definition begins with humans and is extended through domestic animals who are observed engaging in an activity very similar in almost every respect. This same activity is then extended to other creatures whose habits do differ in many respects from Man. If one takes a category such as the ability to respond to sensory stimulus and defines that as sleep, then one has an operational definition capable of testing on any organism. If one stimulates the organism at regular intervals, and if the biologist notices that at one time the organism reacts at one level, while at another period the organism reacts at a markedly lower level, then by the operational definition, the organism is asleep. But how far does such an operational test [which is still practiced in marine biology today] allow one to go? The temptation is to test for one thing and then claim another. One tests for impaired reaction to sensory stimulus and can probably infer from this a depression of physiological functions, the central nervous system and the brain (Aristotle, of course, had no understanding of the brain nor the central nervous system whatever. The functioning of the sensory soul would be the equivalent for Aristotle here.) But one cannot infer unconsciousness or dreaming in fish on the basis of this operational definition. Aristotle accounts for these contingencies in the Parva naturalia. In On Sleeping and Waking 455b 8, Aristotle says:

> But sleep results when operational incapacity occurs not in any chance sense organ, nor for any haphazard cause, but when, as we have just explained, it resides in the primary organ by which one perceives all things. When this is incapacitated, all the sense organs must lose their capacity for sensation; but when one of the latter is incapacitated, the former need not be so affected.

Whatever the controlling soul is, that is what is incapacitated. In fish this is the sensitive soul; hence, the operational definition above is appropriate. The question of dreaming also is satisfied since dreaming is an aberration of the rational soul which fish do not have (On Dreams, 458b 9).

The test for sleeping is most interesting because here Aristotle is defining a process with respect to certain phenomena which can be physically measured. One does not have a controlling a priori theory which causes one to alter observations as was the case with sea urchins. In the case of sleeping fish Aristotle is defining his terms only as far as he can measure, i.e., his conclusions imply no inscrutable events which have not been shown, such as dreaming.{39}

Aristotle begins his discussion about egg laying with the observation that some eggs expand after they are laid.{40}

> First of all, the egg-laying animals have different ways of laying their eggs. a) Some creatures' eggs are imperfect when laid--e.g., those of fishes, which become perfected, i.e., grow outside the creature which produces them. (G.A. 718b 7-10).{41}

This observation, like many others by Aristotle, anticipated an important modern distinction and was not recognized as significant for many centuries. The cleidoic egg has everything it needs to produce an offspring and does not have direct contact with its environment. This type of egg Aristotle calls "perfect" because it does not change in size. The non-cleidoic or imperfect egg is so named because of the absorption of water into the egg along with other trace minerals which increase the size of the egg, while the trace minerals aid in development.

Why are there imperfect eggs? "The reason is that these animals are very prolific and this is their function" (718b 11). Fish by nature must be prolific. In telonomic language one would say fish must be prolific in order to survive as a group since their mortality rate is so high. Aristotle's teleology applies to the group here (as it did in the moth example in Chapter Three). This is justified by considering the population as a higher level unity for (modern) purposes of evolution and (Aristotelian) purposes of positing the eidos of a species.{42} Aristotle's end is perhaps rationally oriented, but it is easily confirmable in population language. The corresponding ME follows this TE:

> . . . if they brought the eggs to a state of perfection inside their bodies, the eggs would of necessity be few in number, whereas in actual fact they produce so many that each uterus seems to be just one mass of eggs. (718b 12-14).

In order to maintain the existence of the species, a fish must lay many eggs.{43} This is the teleological account. In order to lay the maximum number of eggs, the eggs should be as small as possible so that in the limited space of the fish's uterus more eggs may fit. After they are laid and fertilized, then the eggs expand to the size needed for giving birth. The reason the eggs are small is that the physical constraints of space in the uterus combined with the requisite number of eggs to be laid demand a smaller egg. If there is a limit to egg size for giving birth (i.e., an optimum size which is larger than this other size), the only course open is an expanding egg. The entire fulfilling of the telos is conditionally necessary. That is, in order to reach that telos given by the parameters of the system (many eggs in a small space), a certain process must occur producing an expanding egg, viz., one that will satisfy the telos and fulfill the physical restrictions of space. Aristotle explains the achievement of an end through a mechanical process. The case of the imperfect egg demonstrates how ME and TE work together to offer a complete explanation.

The final example in the group of marine animals to be discussed concerns the Cetacea, and specifically the dolphin. These animals give live birth and breathe air. Aristotle recognizes both of these features. He describes copulation at H.A. 540b 22 and birth at 566b. At 589a 32 Aristotle discusses the question of the dolphin's breathing and how this bears on the problem of classification.

> The dolphin is equipped in the most remarkable way of all animals. The dolphin, whale and other similar aquatic animals [Cetaceans] are furnished with a blow-hole. One can hardly allow that such an animal is terrestrial and terrestrial only, or aquatic and aquatic only, if by terrestrial we mean an animal that inhales air, and if by aquatic we mean an animal that takes in water. For the fact is that the dolphin performs both of these processes: he takes in water and discharges it by his blow-hole, and he also inhales air into his lungs; for, by the way, the creature is furnished with this organ and respires thereby, and accordingly, when caught in the nets, he is quickly suffocated for lack of air. He can also live for a considerable while out of the water, but all this while he keeps up a dull moaning sound corresponding to the noise made by air-breathing animals in general, furthermore; when sleeping the animal keeps his nose above water, and he does so that he may breathe air. Now it would be unreasonable to assign one and the same clas of animals to both categories, terrestrial and aquatic, seeing that these categories are more or less exclusive of one another; we must accordingly supplement our definition of the term 'aquatic' or 'marine.' For the fact is, some aquatic animals take in water and discharge it again for the same reason that leads air breathing animals to inhale air: in other words, with the object of cooling the blood. Others take in water as incidental to their mode of feeding. (H.A. 589a 34-589b 16).

Aristotle assesses the problem of classification here. Dolphins breathe air for the sake of cooling their blood in the same way that land animals do. Thus it seems that in this respect they are similar to land animals. But their habitat is the water which is in common with marine animals. Dolphins are unlike most marine animals such as fishes, who take in water to cool their blood. Dolphins take in water, but not to cool their blood. It has more to do with digestion than respiration. Thus the dolphin seems to be in a class with the whale who also seems to be in an intermediate classe. However, if one is to emphasize attributes over habitat, as is suggested in P.A. 1 (643b 10), then dolphins belong more with the land creatures. Because of their respiratory and reproductive behavior they seem less akin to fish than to land animals. But with respect to their appendages and skin they seem far removed from the land animals. Perhaps it is best to classify the Cetacea as somewhere between the two groups.{44}

It is important to observe that as Aristotle approaches this problem he does not search for some overarching a priori principle which would once and for all decide the question. Instead Aristotle examines many differentiae in an effort to see if collectively they will point to one side or the other. By doing this Aristotle is acting in accord with his stated ideas concerning classification. Dichotomy would place the decision into the power of one differentia. Aristotle's pluralism requires one to consider all the relevant observational data. This is not to say that each characteristic is viewed equally, since organized systems (such as respiration) seem more important than isolated parts, but rather taken as a whole, many differentiae will offer a more complete picture than just one will.

4.4 BIRDS

In Aristotle's treatment of birds one finds a strange mixture of experimentation and a priorism; critical scrutiny of sources and almost blind acceptance of others. These inconsistencies lead one to believe that Aristotle's own observations on birds were more limited than his extensive work on marine animals. One of the most frequently examined passages in Aristotle's biology concerns his famous work with developing eggs. In the Historia animalium Aristotle says, "Birds without exception lay eggs, but the pairing season and the times of parturition are not alike for all" (588b 10). He goes on to describe the habits of several domestic varieties which he undoubtedly observed first-hand.{45} The eggs are physically described. The exposition seeks no explanation, only description. The reason for the "perfect" egg is delayed for the De generatione animalium. Aristotle explains that an egg is perfect due to its heat as he argues for a scale of animal birth from viviparous to the two varieties of oviparous (G.A. 733b). There is an unexplained link implied between an animal's heat and the place on this hierarchy of procreation. But perhaps of more philosophical interest is the description of embryonic development. Aristotle asserts that the embryo is not a preformed being which only grows bigger through time, but a potentiality imposed upon matter that develops one part at a time.

> I mean, for instance, not that the heart once formed, fashions the liver, and then the liver fashions something else; but that the one is formed after the other (just as

> man is formed in time after a child), not by it. The reason of this is that so far as the things formed by nature or by human art are concerned, the formation of that which is potentially is brought about by that which is in actuality; so that the form of B would have to be contained in A, e.g., the form of liver would have to be in the heart—which is absurd. (G.A. 734a 28-35).

This process is likened to a chain of reactions and causes. After the heart is formed, the liver can be formed and so on. The logos is in the zygote as potentiality. As it develops into an embryo, certain aspects of this potentiality actuate themselves. It is not as if they are actually there, as preformation would assert, but that the parts can develop from a potentiality. In modern terms, one might say that in order to arrive at such and such end, such and such genetic information is needed so that the organism might epigenetically develop. Though there may be an end to the process, this does not mean that anything is "preformed." For preformation calls for all the parts to be formed [or to be present] at once. Epigenesis calls for a determined hierarchy in formation of body parts. When this process is controlled according to a plan (as is thought to be the case in modern as well as in Aristotelian biology), then the whole process can be explained teleologically. Mechanically, one understands the event as analogous to the wind-up puppets Aristotle describes. In generation the sperm and ovum unite through an external agency. After this, no external agency plays a part in an embryo's development since the embryo is enclosed and no external forces are in "contact" with it (G.A. 734a 5).{46} Thus the embryo develops on its own after having been "set in motion" by sexual union. The whole process should be understandable in mechanical terms. It is the same with these wind-up puppets which, after being set in motion, appear to go by themselves.{47} Just as the fertilized egg, the puppet has been set in motion by an external force. Also just like the puppet, the fertilized egg continues by itself. The whole process can be explained as a nexus of causal interactions in time.

> And it is possible that A should move B, and B move C, and that the process should be like that of the "miraculous" automatic puppets: the parts of these automatons, even while at rest, have in them somehow or other a potentiality, and when some external agency sets the first part in movement, then immediately the adjacent part comes to be in actuality. The cases then are parallel: just as with the automaton in one way it is the

> external agency which is causing the thing's movement—viz., not by being in contact with it anywhere now, but by having at one time been in contact with it. (G.A. 734b 10-18, cf. 741b 9).

This passage shows how well explanations of necessity can work with teleological ones. To further support his point Aristotle actually looked at eggs as they were developing and noticed the heart developing first (P.A. 665a 35; see 4.7).

Further Aristotle tried to identify various stages in the process through examining the role and function of the white and yoke of bird's eggs (G.A. 751a 31). Aristotle says that one part of the egg is hotter, the white part, since it is here where the soul of the bird will develop (751b 4, cf. 762a 18, P.A. 652b 7).{48} The <u>archē</u> of the embryo is obtained through the white.

> That is why in the case of all double-colored eggs the young animal gets its "principle" of generation from the white, because hot substance is the place where the soul-principle is to be found, while it gets its nourishment from the yolk. With those animals, therefore, whose nature tends to be hotter we find there is a clear distinction between the part from which "principle" is formed and the part from which the nourishment is derived: the one is white, the other yellow. There is always more of the pure, white part than there is of the earthy, yellow part. With the animals that are less hot and more fluid, there is more yolk in the egg and the yolk is more fluid. (G.A. 751b 5-15).

Aristotle's argument seems to be:

First Argument

1. Heat is akin to the <u>archē</u> (i.e., soul and life).

2. The white is hotter than the yollow.

3. The white must be the <u>archē</u> of the egg.

Second Argument

1. The yellow is heavy and tends to overcome the white.

2. The hotter the white the more the two tend to separate.

3. In colder animals the white is colder and tends to mix with the yellow.

Concerning the first argument, Aristotle's first premise is supported by the fact that he makes the heart the arche of an organism. The heart makes the body hot, for when it stops (in death), the body becomes cold. This argument is asserting a broad causal connection, not completely justified, between the heart and heat. It is true that when the heart stops, the body becomes cold, and that the heart's stopping may be the cause of death (though some want to say that it is the brain), but that does not mean that the heart makes the body warm. It is a necessary condition for the metabolic activity of the body to transpire, but it only makes it possible for the real heat producers (those involved with respiration, the creation and breakdown of the ADP and ATP molecules) to do their work. Aristotle's mistake occurs because he does not have an adequate mechanical account of the production of heat. He seems to think the solution is obvious, and so he makes several large causal connections which he refrains from doing on points he considers to be more controversial.

The second premise that the white is hotter than the yoke is stated with no justification. Did Aristotle try to feel each? I think rather that this point is assumed because it fits in with his observation that the chick appears around a yoke sac, and that naturally the chick must have been formed from the white.{49} If the white is the arche of the egg, it must have been hotter. The 'must' is a logical necessity. Again Aristotle seems to believe that he has it right, but because of this belief, he does not feel it necessary to check his conclusions with the egg itself. Is one hotter than the other? Would it be testable? Certainly he could try, to the rough degree of sensitivity of which his fingers were capable, to test it. Aristotle knows what he saw, a developing organism formed around a yoke sac. His observation was correct, but his account is not exactly correct because he did not follow his own painstaking method which might have allowed him to avoid a mistake. Neither the yoke nor the white play a simple role in the embryo's nourishment nor in its arche. Much development does take place around a yoke sac though the cicatricula is within the same yoke. Though Aristotle may not have been able to determine this, he could have avoided the mistakes he did make by performing a simple test.

Again in the second argument Aristotle could have observed by cracking open an unfertilized egg that the white has a higher specific gravity than the yoke and so his entire argument about hot and cold does not account for differences observed between the relative mixing of yoke and white as one moves down the hierarchy of animal development.{50}

In both of these examples Aristotle makes interesting and accurate observations of biological processes, but by bringing in a supporting theory derived elsewhere, his explanatory conclusions drawn from his data are not correct. The problem is not his bringing in an outside theory, for as shown in Chapter One such theories are necessary for any explanation whatever. A biologist must be able to bring in a theory to describe novel events. But he must also test the application of that theory at every stage of the process to determine whether the application was justified. Whenever one fails to do so, he is departing from his empirical base and inviting errors in his _explanans_.

One instance in which Aristotle (or a Peripatetic practioner) seems to adopt common opinion in an uncritical way is found in the description of bird aggression. As D'Arcy Thompson documents in his excellent work _Greek Birds_,{51} Aristotle cites approximately forty instances of animal hostility. About half of these are also mentioned by other writers. Most cases of this latter group concern birds.{52} Now many of the examples that Aristotle uses to illustrate hostility come from myth or are depictions of astrological events. Examples abound such as the enmity between the eagle and the hare (H.A. 618b) which is represented on gems and on coins, indicating a popular mythological usage (Thompson, p. 110). The owl and the crow are declared to be enemies (609a 9). This story is documented as an ancient oriental tale in the _Mahabharata_.{53} The account as it appears in Julian, _Imp.Orat. 4, 149_, suggests that the story is a parable of the sun and the moon. Thompson elaborates on the astrological analogues:

> Many illustrations of this theory of mine will be found in the pages of this Glossary. Suffice it to say here, in briefest illustration, that the Eagle which attacks the Swan and is in turn defeated by it, is, according to my view, the constellation Aquilla, which rises in the East immediately after Cygnus, but, setting in the West, goes down a little while before that more northern constellation; that Haliaetus and Ciris are the Sun and Moon in opposition, which rise and set alternately, like the opposite constellations of Scorpio and Orion with

which the poet compares them. (page xiii){54}

If these statements are correct then one is tempted to reject the whole lot of these conclusions from the zoological canons. Clearly these accounts have nothing to do with phenomena-1 at all. These stories are common opinion which, because of their very source, deserve verification. Accepting the common opinion of a group of beekeepers about bees is one thing, but accepting accounts about bird aggression from poets and ancient fables and/or astrologers is quite another thing.{55} All opinions are not of equal weight. We do make distinctions concerning the veracity of our sources.{56} Some sources are deemed accurate and others inaccurate. These epitaphs are applied partly due to the method of acquisition and partly to acquaintance with the subject matter. The practioner of these passages has accepted a body of information which is acquired in a manner contrary to the method outlined in P.A. 1. To be fair to Aristotle, however, it should be noted that the ninth book of the Historia animalium, in which these passages occur, has been argued by some to be a Peripatetic work and not authentically Aristotelian.{47} Be that as it may, the method employed in these passages on bird aggression is not in accord with the canons of P.A. 1.

When Aristotle describes bird song he follows his method much better. To begin, Aristotle differentiates 'sound' from 'voice' (H.A. 535a 27). Sound is any noise made by the animal such as the buzzing of a bee. Voice comes from wind passing through the pharynx (P.A. 664a 36, H.A. 535a 28).{58} When tones are emitted by voice these tones take shape and go into the ear thus shaped (De anima 419b 3, cf. De sensu 446b 5). When the purpose is comminication, then language is the result. Thus, there are two categories: sound and voice. Voice, in turn, can be in language or not in language (H.A. 535a 27). With this framework in mind, Aristotle has to account for the mechanical reason why some animals can have language and others cannot. Aristotle believes that this can be explained by the tongue.

> A tongue which is broad can also become narrow, on the principle that the great includes the small, but not vice versa. That is why the clearest talkers, even among birds, are those which have the broadest tongues. On the other hand, the blooded viviparous quadrupeds have a limited vocal articulation; it is because their tongues are hard and thick and not sufficiently loose. (P.A. 660a 27-32)

THE PRACTICE

The three properties of tongues are: breadth (cf. H.A. 540b 1), thickness and looseness. A thick or attached tongue cannot move around to articulate the undifferentiated noise of the air which has passed through the larynx. To do this, various configurations are necessary. The more configurations a tongue can make, the greater the mechanical capacity to make a large variety of voices. Thus the tongue best suited for being formed in many shapes is a broad, thin and loose one. Aristotle's considerations here are strictly mechanical via applied geometry. But what is the TE for going through all of these configurations and making such a large variety of voices?

> All birds use their tongues as a means of communication with other birds, and some to a very considerable extent; so much so that it is probable that in some cases information is actually conveyed from one bird to another. (P.A. 66-0a 35-66b 2).

Language or communication is suggested by Aristotle to be possible for birds, but at H.A. 536b 1, Aristotle says, "This power, or language is peculiar to Man." A few lines later Aristotle qualifies this statement.

> Vocal sounds and modes of language differ according to locality. Vocal sounds are characterized chiefly by their pitch, whether high or low, and the kinds of sound capable of being produced are identical within the limits of one and the same species; but articulate sound, that one might reasonably designate 'language,' differs both in various animals, and also in the same species according to diversity of locality. (H.A. 536b 9-14).

Aristotle's point here is that Man has the most significant form of communication among animals. Compared to Man the only animal that comes close to communicating anything with its voice is the bird. But even if the bird does communicate, it is on a very rudimentary level compared to Man's linguistic capacity.

I feel confident that Aristotle would allow several different types of vocal communication in other animals besides birds. For example, alarm calls are not infrequent in social animals which tend to flee from their predators. Again such primitive degrees of communication are limited to a few sign-stimulus-responses, whereas language calls for thousands.

THE PRACTICE

At any rate Aristotle believes bird calls are not merely idle singing, but for the purpose of communicating. This teleological account is supported by observations of specific instances of communication.

> The smaller birds are more vocal and given to chirping than the larger ones; but in the pairing season every species of bird becomes particularly vocal. Some of them call when fighting, as the quail, others cry or crow when challenging to combat, as the partridge, or when victorious, as the barn-door cock. In some cases cock-birds and hens sing alike, as is observed in the nightingale, only that the hen stops singing when brooding or rearing her young; in other birds, the cocks sing more than the hens; in fact, with barn-door fowls and quails, the cock sings and the hen does not. (536a 22-32).

Birds sing in mating and territoriality (fighting in the pre-mating season). Also females sing to their developing young. These are specific functions which bird song might have as a communicative tool. When observed in these activities the song seems to play some definite role. In territoriality it is keeping another bird out of his potential mating area. In mating it is choosing the most desireable female (a function of the song and the territory, among other factors). And in rearing young the function is unclear unless it is believed that the mother is "teaching" her brood a particular song. This is exactly what Aristotle is suggesting.

> Of little birds, some sing a different note from the parent birds, if they have been removed from the nest and have heard other birds singing; and a mother-nightingale has been observed to give lessons in singing to a young bird, from which spectacle we might obviously infer that the song of the bird was not congenital with mere voice, but was something capable of modification and of improvement. Men have the same voice or vocal sounds, but they differ from one another in speech or language. (536b 14-20).{59}

Just as particular accents and the like vary according to the locality and are passed on by men, so also birds pass on the idiosyncrasies of their flock. To prove this Aristotle says that the song of a young bird will alter when removed from its mother at birth. This is the correct test for bird song origins and has been used by W.A. Thorpe to indicate that with Chaffinches there are two parts to a bird's song, the full song and the sub-song.{60} The full song consists of well-defined bursts of sound between two and three seconds, while the sub-song is an indefinite series of irregular notes continuing for a half minute or so. The full song is derived from the sub-song by the Chaffinch as it matures. It discards the irrelevant notes as it listens to other chaffinches. Now a Chaffinch which is deafened before acoustical feedback is possible will develop a similar song (from his sub-song) to that of his natural group, but the song will have some small variations from those songs of non-deafened members of his natural group. One explanation for this that Thorpe and others give, is that much of the process of making a song from a sub-song is somehow coded genetically and expressed in the central nervous system of the bird.{61} If this account is true, then at least some or most of the Chaffinch's song is innate. But even though much of this behavior is determined, still there is a significant portion which is acquired in the bird's environment by listening to other birds around it sing. Thus Thorpe concludes, from essentially the same kind of evidence that Aristotle had, that there is a part of a bird's song (in some species) which is learned.

Whether Aristotle actually carried out this experiment or not is impossible to know. But I believe that if Aristotle did not actually do it himself, he knew someone who did, since there is no a priori mechanism which would give him this answer. If anything, it would have been easier to say carte blanche that only Man communicates through vocal language. Aristotle's statement that not only do birds communicate in territoriality and mating, but also that this song is taught to the young is so specific that it seems as if it must have been the result of an experiment.

4.5 DIFFICULT CASES

This short section will deal with those cases which do not fit neatly into Aristotles systematics, and the way Aristotle handles such cases. These different cases fall into two groups: 1. Accounts Aristotle has received which talk about a strange animal or which attribute incredible traits to a known animal, and 2. Animals who do not fit into only one class.

Aristotle has differing attitudes about these stories. At times he seems to believe them such as H.A. 532b 18 where Aristotle says:

> Furthermore, there are some strange creatures to be found in the sea. Because of their variety we cannot classify them. Experienced fishermen affirm they have sometimes seen sea animals like sticks: black, rounded and of the same thickness throughout. Others claim that they have seen creatures resembling shields, red in color, and furnished with fins packed close together; and others that they have seen creatures resembling the male organ in shape and size, and with a pair of fins in the place of the testicles. They claim that on one occasion a creature of this description was brought up on the end of a nightline.

These creatures still cannot be identified from the account given.{62} Aristotle does not indicate his view except to say that there are strange creatures and then to identify the source.

A similar approach is taken at 631a when Aristotle relates some strange feats attributed to the dolphin.

> Incredible stories are told about the rapidity of movement of this creature. It appears to be the fleetest of all animals, marine and terrestrial. It can leap over the masts of large vessels. This speed is chiefly manifested when they are pursuing a fish for food; then, if the fish endeavors to escape, they pursue him in their ravenous hunger down to deep waters. When the necessary return swim is getting too long, they hold in their breath, as though calculating the length of it, and then draw themselves together for an effort and shoot up like arrows, trying to make the long ascent rapidly in order to breathe. In the effort they spring right over a ship's masts if a ship be in the vicinity. This same phenomenon is described in divers, when they have plunged into deep water; that is, they pull themselves together and rise with a speed proportional to their strength.

Aristotle appears to be offering TE and ME for a dolphin's speed in the water and its ability to jump over a mast of a ship.{63} Here again, as with the case of aggression among birds, Aristotle fails to critically evaluate his material. Certainly a diver breaks the surface of the water when rising quickly, but this does not give credence to the story. As noted [63] a dolphin would have to travel at fantastic speeds in order to leap over the mast of a tall ship. Surely this premise of the argument should be viewed by Aristotle more skeptically. There are other reports scattered throughout the *Historia* which are of this nature reporting such amazing creatures as dragons (621a 30).{64} But there are other contexts in which Aristotle seems less inclined to believe some of these (from our perspective) far fetched tales. For example, in the De generatione animalium Aristotle attacks the view that fish may become impregnated by swallowing milt.

> The fishermen do not notice this, but they do notice the swallowing of milt and eggs by the female, and so they join the chorus and repeat the same old stupid tale that we find told by Herodotus the fable-teller, to the effect that fish conceive by swallowing milt. It never occurs to them that this is impossible, but of course it is, because the passage whose entrance is through the mouth passed down into the stomach, not into the uterus, and whatever goes down into the stomach must of necessity be turned into nourishment, because it undergoes concoction. The uterus, however, as we can see is full of eggs; so we ask, how did they find their way there? (756b 5-12).

Here Aristotle pokes fun at the "mythology" of Herodotus (2.93) by showing that such an account violates what is known about the specialization of internal systems. The refutation follows from an analysis of the theories of digestion and reproduction which had been accepted via dissection and observation. If Aristotle could be critical of naive acceptance of stories in the De generatione animalium how could he have accepted so many of these tales himself in the Historia animalium? Perhaps H.A. is only meant to be a notebook and not critical or evaluative. In this case, Aristotle would just be recording facts, but not examining them. This is perhaps partially true, but the last case of bird song is an instance of Aristotle evaluating cases in the Historia. Perhaps Aristotle had not made his mind up yet about these stories when he wrote H.A. This may be true, but it fails to account for his selection of stories elsewhere. What makes the fish milt account different from the jumping dolphin account is that in the former case the event is physically impossible because of Aristotle's understanding of physical processes. The latter event could happen if the dolphin could swim fast enough. One is logically impossible while the other is not. For this reason Aristotle rules the one out of court and reports the other as a possibility (though its confirmation rests with others).

In the case of animals that are in between groups, Aristotle says that they tend epamphoterizein (to intermediate).{65} The animals which do not fit into any one group are those with attributes of more than one group. In deciding whether or not a sea anemone is a plant or an animal Aristotle says:

> In their nature they incline towards the plants on one side and the animals on the other. Towards the animals, because some of them detach themselves and fasten upon their food, and are sensible of objects that come up against them; and also because they make use of the roughness of their body for self-preservation. Towards the plants, because they are incomplete, and quickly attach themselves to rocks; and further, because they have no residue that can be seen, though they have a mouth. (P.A. 681b 1-9).

The sea anemone appears to have a double source in resembling aspects of plants and animals. How does one decide to classify it? Obviously one must determine a "crucial" characteristic, but what if this organism shares two crucial traits?{65}

The problem comes up again with the seal which seems to be both a land and a sea animal.

> Similarly, seals and bats are in an intermediate position. Seals are between land-animals and water-animals, bats between land-animals and fliers: thus they belong to both classes or to neither. Seals, if regarded as water-animals, are anomalous in having feet; if regarded as land-animals, in having fins (their hind feet are altogether like those of fishes—i.e., fins; and all their teeth too are sharp and interlocking. (P.A. 697b 1-7).

Seals can go on land, breathe air and bear live offspring as land quadrupeds, but they have fins, spend most of their time in the water and have saw teeth like the fish. Both in habitat and anatomical structure the seal is depicted as an intermediate.{67}

One could view such admissions as problems in Aristotle's systematics [and certainly modern theories do much better here, see: 2.5] or see them as instances of one continuous <u>scala naturae</u>. But either of these two readings misses Aristotle's <u>actual</u> point here which is to show how varied the phenomena are. When one tries to break animals into natural groups, there are organisms which do not fit so neatly. To put these animals into one group or the other may artifically cut them off from those properties they share with another group. Aristotle seems to separate these difficult cases into their own group, the intermediates. The presence of such a group should indicate to one that the complexity of biological organisms is tremendous and that it is precisely because of these intermediates that generalizations in the biological sphere must be made very carefully over a specified class. Because of the nature of the *explanandum* there are numerous instances in which a biologist may overstep in his eagerness to set down universals. The many differentiae of biological organisms thus affects not only classification, but the whole manner of biological explanation (a fact which is dramatically emphasized by these difficult cases).

4.6 UNIFORM PARTS

The uniform parts include flesh, blood, lard, suet, marrow, semen, bile, milk (P.A. 647b 13) and the brain (652a 25). The uniform parts are by nature homogeneous (647b 18). They may be divided again and again. This is different from the non-uniform parts which cannot be divided. For example one may divide a cup of blood many times and still have blood, but if one divides a face once, he no longer has a face. The relation between the two kinds of parts is stated in three ways:

> Now first of all there are many sorts of cause to which the existence of these uniform parts, both the fluid and the solid ones is to be ascribed. Some of them act as the material for the non-uniform parts. Another group of the uniform parts—the fluid ones—act as nutriment for the ones just mentioned, since everything that grows gets the material for its growth from what is fluid. (647b 22-28).

Bone, which is a uniform part creates the leg (an instrumental part and non-uniform) which is composed of specific bones like the femur (a non-uniform part). Blood is the chief example of the second group and various types of residue (*perittoma*) include marrow, semen, milk and lard.

One of the most important uniform parts is flesh.

> Let us first take flesh (and where flesh is absent, its counterpart). This is both a principle and a body in itself. Its primacy can also be logically shown as

> follows. We define an animal as something that has the power of sensation, and chiefly the primary sensation, which is touch; and the organ through which this sensation is effected is the flesh (or its counterpart). And flesh is either its primary organ (comparable to the pupil in the case of sight), or else it is the organ and the medium of the sensation combined in one (comparable to the pupil plus the whole of the transparent medium in the case of sight). Now not only was it pointless, it was impossible for nature to make such a combination in the case of the other senses . . . It is also clear from our actual experience in sensation that all the other parts exist for the sake of the organ of touch (the flesh). In these I include the bones, the skin, the sinews, the blood-vessels; also the hair, nails of every sort and kind, and the like. (653b 20-33).

If 'animal' is defined as an organism having the power of sensation [sensitive soul], and if sensation is impossible without flesh, then flesh is primary in our definition of an animal. In sensation, one has the receptor, the medium and the object. In sight, the receptor is the lens of the eye. The medium is the air between the object and the receptor [Figure 4—(a)].

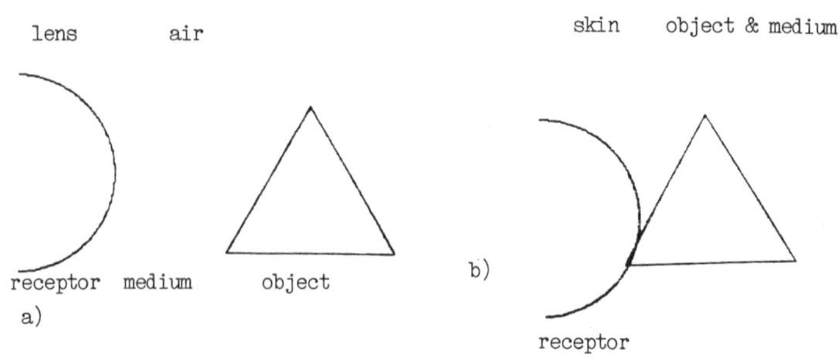

FIGURE FOUR: Skin and Sensation

In the case of touch [Figure 4—(b)], the medium becomes the object itself (cf. *De anima* 423b 5). Thus touch is the most corporeal of the senses. In addition to this argument which is essentially a rational argument, Aristotle offers another argument. This one is prefaced by the words, "it is clear from our actual experience . . . " With such a beginning one would expect this argument to be empirical, but it is not. Aristotle offers a teleological argument that because the skin is the outermost covering (which we can see), the interior parts must exist for the sake of the skin. But this is a queer argument.

It is analogous to saying that the engine of an automobile and all the internal parts exist for the sake of the metal body. Why not say the internal organs exist for the sake of the whole? What is more primary about the skin over say the blood vessels under the skin? This so-called empirical argument is nothing more than another _a priori_ argument. Aristotle wants to assert that the hard parts exist for the sake of the soft ones, but he gives no empirical grounds for this nor even a convincing rational one.

Aristotle does comment on the softness of human flesh (660a 10). He tries to draw a mechanical connection between the softness of skin and the sensitivity of touch. Having no understanding of nerves, Aristotle makes a reasonable assumption since a calloused finger is less sensitive than an uncalloused one. Aristotle is undoubtedly making an extension from human hands (or feet) to all animals. This statement also follows from his statement on contact, since a calloused hand represents a medium (the callous) between the finger and the object. Any added medium will dull a sense, hence; tough skin is less sensitive than soft skin.

Though Aristotle does notice this feature about skin, he fails to recognize the special nature of skin, its elasticity. Nowhere is this property noted or explained. There are, however, passages where properties are assigned to flesh, such as breaking into circular portions (654a 17). This statement came perhaps from examining fingerprints or a butcher with a unique cutting technique or some other remote source.

Though Aristotle gives flesh a high place in the process of sensation, he makes it clear in _De anima_ that the actual sensing does not take place in the skin.{68}

> . . . the tactual medium through which the several sensations are felt must be an organically attached body. That they are several is clear from a consideration of touch in the case of the tongue; for the tongue perceives all tangible objects with the same part with which it perceives flavor. If then the rest of the flesh also could perceive flavor, taste and touch would seem to be one and the same sense. But, as it is, they are proved two, because they are not convertible. (_De anima_ 423a 16-21).

It is from this passage that touch and flesh are not the same. Indeed, when one thinks about 'touch' he includes not only sensations, but also sensations as organized. This organization cannot occur in the skin Aristotle asserts:

> It must evidently be by some sense that we perceive the difference; for they are objects of sense. Incidentally it becomes clear that flesh is not the ultimate sense organ; for, if it were, judgment would depend on being in contact. Nor, again, is it possible to judge that sweet and white are different by separate sense, but both must be clearly presented to a single sense. (*De anima*).

It is touch (*aphēs*) not flesh (*sarkos*) which is declared to be primary. Touch is what makes the sensitive soul possible for it is the power of sensation.

> Without touch there can be no other sense, and the organ of touch is composed neither of earth nor of any other single element. It is obvious, then, that deprived of this one sense alone, animals must die. (*De anima*).

When Aristotle wrote *De partibus animalium* he may not have had his explantion of sensation entirely worked out. In that work flesh seems to play the primary role in the process of sensation. But a common sense was needed to organize what would remain essentially chaotic if flesh alone were involved in sensation. This organizing element is the power of the soul for touch. This power is the capacity to receive data from the senses and organize it into a coherent whole.

Where does this leave flesh? Perhaps flesh is relegated to the place of medium. Whereas the medium in P.A. 653b 20 seems to be absent, in *De anima* Aristotle declares, "the medium of the tangible is flesh" (423b 26). Though this appears to be a step backward from the exalted role for flesh declared in P.A., Aristotle's later view seems better able to account for a wider range of phenomena than his earlier view. That earlier view was characterized by *a priori* arguments and Aristotle must have felt some dissatisfaction with them. So it is that his arguments about touch in *De anima* feature more empirical examples.{69}

Perhaps of even more importance to the organism is the uniform part, blood. Blood provides nourishment to the body (647b 27) and plays a significant role (i.e., material cause, cf. 651a 12) in the formation of all other body parts (G.A. 726b 5). Aristotle has blood produced in the heart which holds blood "like water in a jar" (P.A. 650b 10). From there the blood passes nourishment to the body from very small vessels that are practically (though never completely) continuous with the skin. The two parts of blood are the watery part and the earthy part (650b 19). The watery part or serum (*ichor* 651a) is the principal constituent and is cool, while the fibers (*hines*, H.A. 3, 6), which allow blood to coagulate, are hot.{70} Aristotle observes that blood in the body is not of uniform color or consistency throughout the body (647b 35). He attributes this to varying amounts of fiber or serum. It is interesting to note that from one correct observation, the difference in blood's consistency and color, Aristotle, because of his explanatory framework, is also able to claim that there must be differences in temperature as well. Now it is true that slight variations in temperature can be observed, but these fluctuations were too slight for an observer without precise thermometers. Aristotle's second statement is possible in virtue of his *explanans* for blood. Blood does appear to the unaided eye to *have the* elements that Aristotle ascribes to it. But by saying that the serum part is water and hence cold and that the fiber is made of earth and therefore hot, he is imputing properties to blood that he has no right to ascribe.{71} Aristotle sees a thin liquid part of blood and another part which is cohesive and seems heavier (this dual character is most noticeable in congealing blood). This Aristotle can see and name. But to go further is not justified. If he could show that thinner blood was cooler or that thicker blood *was* hotter, then his ascription *would* be supported, but as it *stands* he has attributed to the two types of blood he has observed, these qualities, simply because it follows from what he has earlier laid down. This leads Aristotle into several other errors of greater magnitude. First it allows Aristotle to connect it with another *a priori* statement, that the right side of the body is more *noble than* the left,{72} and argues that the right side of the heart (and the whole right side of the body) is hotter than the left side (667a, cf. 648a). In this case Aristotle has taken the bogus conclusion that some blood is hotter than other blood and combined it with his *a priori* predilection for the right over the left.{73} This *is clearly* a statement that Aristotle could have checked, but did not. He must have felt that it was so obvious from his argument that no one could object. He even uses it at G.A. 765b to show semen concocted on the right side of the body is better than that from the left side.{74} Hotter semen coming from the right side will be "set more" and therefore be more fertile (i.e., better). This argument is presented as an ME, but it is based upon the premise that the right side is hotter than the left side of the body, a conclusion which has been shown to be unjustified. One error compounds itself many times. Aristotle could have avoided this at several stages by bothering to verify his conclusions.

A second problem with the premise that thin blood is cool and thick blood is hot can be seen in relation with Aristotle's assertion that thin blood is conducive to intelligence while thick blood is associated with strength (648a 5). Peck correctly identifies this as a usage of a Hippocratic doctrine found in _On Regimen_ 1, 35.{75} In that treatise the argument is made that the coolest heat and the dryest water are conducive of _phronēsis_. This is because they are sulf-sufficient and able to exist at the extremes. The cool, wet is then generalized to be beneficial for a number of medical needs because of its power of "intelligence or sensitivity" (_phronēsis_). Aristotle uses the same word and seems to be borrowing the connotation as well. However, several problems occur when Aristotle applies this to organisms. He wants to claim that it is this quality of thin, cool blood that distinguishes certain animals as being more intelligent than others (650b 20). This even applies to the analogous substance found in the bloodless animals. Bees and ants, for example, have a more mobile faculty of sensation because of the thinness of their"blood" (i.e., its analogue) which makes them more intelligent than even some blooded animals (650b 24). Aristotle, it can be remembered, during the dissection sections of insects in the _Historia animalium_, does not make the observation that bees or ants have remarkably thin body fluid. This conclusion seems to follow directly from his own distinction that blood can be thin or thick and correspondingly is associated with the Hippocratic adage about thin blood and intelligence. Aristotle noticed that bees and ants have remarkably complex social behavior. To what can this be materially attributed--i.e., what is the ME for this? Aristotle conveniently offers the thin-blood-intelligence theory. His conclusion is not based upon any observation, but merely on the concurrence of a hasty conclusion of his own and a common opinion.

To make the explanation seem even more convincing, Aristotle declares that animals that are observed to be easily angered, such as bulls and boars, have an excess of the thick, easily congealing blood. Again Aristotle has simply fit together a trait with a handy theory. He could have easily checked it, but it is clear that he did not.{76} Aristotle did not bother to verify his conclusion because it was not empirically derived. It is often the case with Aristotle that those theories which are empirically derived are empirically checked. Those which are rationally derived are only checked rationally. But even here this last theory seems to be in trouble when one considers that the center of sensation which is around the heart is also the hottest spot in the body. If the heart is the center of the sensitive and rational soul, then one would expect the blood to be cold here. However, the 'cold' aspect seems to be more in evidence in the 648a 5 passage than in the 650b 24 passage which emphasizes thinness over coldness. The former pasage is definitely problematic while the latter is capable of being modified to fit Aristotle's conception of the heart. This non-uniform presentation may be just another indication that Aristotle is toying with how he wants to use this Hippocratic common adage.

These various examples of scientific argument illustrate that though a theory may offer a material/mechanical explanation, it may still be _a priori_ in nature. Aristotle's principal problem, in these explanations generated from the thin and thick aspects of blood, lies not so much in his ascription of earth and water to the fiber and serum parts of the blood, but rather in his inclination not to test his results. Several times empirical verification was open to Aristotle which he does not employ as he did so often when writing about insects and marine animals. This may suggest to some that Aristotle is inconsistent in practicing his announced theory, but I tend to think that the real trouble stems from the social prohibitions which prevented Aristotle from engaging in human dissections.{77} Because of this Aristotle was forced to work by analogy and his best guesses were often the result of such rationally based explanations as have been presented above.

Nowhere, perhaps, in Aristotle's work on the human body are these limitations more evident than in his work on the brain. At the time of Aristotle's writing there were several opinions about where the center of sensation and thought were. The Hippocratics put this center in the brain. Aristotle rejects this by the following argument.

1. The brain has no blood vessels in it—observation (H.A. 514a 15).

2. Blood only travels in blood vessels—observation.

3. The brain has no blood--from 1 & 2.

4. Thin blood causes sensation, i.e., is necessary for sensation—conclusion from the _a priori_ argument above (cf. 656b).

5. The brain has no sensation--from 3 & 4.

Premise one is difficult to understand. In almost every mammal blood vessels are quite easily seen within the brain itself. Aristotle, however, believed that blood vessels only surrounded the brain in a delicate membrane.

> The brain itself in all animals is destitute of blood, and no vein, great or small, holds its course therein. But of the remaining veins that branch off from the last-mentioned vein some envelop the head, others close their courses in the organs of sense and at the roots of the teeth in veins exceedingly fine and minute. (H.A. 514a 16-22).

The observation of a membrane around the brain demanded perhaps even a closer observation than noticing the existence of blood vessels in the brain. Lones suggests that this latter phenomenon can be accounted for if Aristotle only observed a cooked brain (perhaps a course at dinner?).{78} In a cooked brain Aristotle would not notice any blood vessels. However, if this were true, then the observation about the membrane around the brain must not be Aristotle's or if it is, it must have been made on a fish or reptile.

Aristotle's first premise, if it is based upon a first-hand observation, is not based upon a live mammal or a dead one which would yield a more accurate comparison than a remote species such as a fish. If one accepts his inaccurate first premise and the correctly observed second premise and the <u>a priori</u> fourth premise, the conclusion follows well enough. This whole argument is a good example of the mixture between good observation, bad observation and naively accepted common opinion. The form of the argument is good: a mixture of observation and scientific opinion, but the accuracy of each leaves something to be desired. Aristotle is following his method formally here by employing empiricism and ME, but it is the <u>manner</u> in which he follows it that is important. He is partially fulfilling the letter but not the spirit of his empirical method when he only observes (assuming this suggestion is true) a cooked brain or a brain of a genus far different from Man to describe Man. His acceptance of the Hippocratic doctrine of the thin-blood-sensation theory ahs already been questioned earlier and Aristotle's use of the premise here is subject to the same questions. Aristotle's conclusion is wildly wrong not because he employed an improper method, but because he employed his method improperly.{79}

As a result of this argument, Aristotle asserts that the function of the brain is to cool the blood by balancing the heat of the heart (652b 22).{80} According to Aristotle, one can tell that the brain is cool by touching it (652a 35). Again one is curious to know what brains and in what state they were when Aristotle touched them. One might touch the flesh of a corpse and say, because it is cold, skin *per se* is cold. Even Aristotle admits that all heat leaves the body at death (*Youth, Old Age, Life and Death* 469b 21) so that to touch a dead brain is hardly empirical support. It is again interesting to note that Aristotle feels the need to offer sense data as support even if it is of ultimately little use. The whole process is compared by Aristotle to the water cycle 653a 5. The heated blood rises to the brain where it is chilled so that it falls. Just like the rain which falls from the clouds only to rise again, so also the blood is said to move about. In this way the phlegm and serum levels are controlled. One has here a form of circulation by convection with the two major organs, the brain and the heart, paired against each other as the center of heat and the center of coolness respectively (652b 22).{81} The brain acts, therefore, for the sake of cooling the blood and making the blood duly proportional in temperature between the heart and itself.

> For this cause Nature has contrived the brain to counterbalance the region of the heart and the heat in it; and that is why animals have a brain, the composition of which is a combination of Water and Earth. Hence, although all blooded animals have a brain, practically none of the other has (unless it just be a counterpart, as in the case of the octopus), for since they lack blood they have but little heat. (P.A. 652b 21-25).

Aristotle offers the TE for why the brain must be cold. He has already gone through an argument on why the heart and not the brain must be the center of sensation, relying primarily on empirical premises and ME. Now he is attempting to show that the brain is cold and serves as a counterbalance to the heart. To illustrate this Aristotle brings out his empirical support of touching the brain as well as the TE of its purpose.

From this account Aristotle is able to attribute sleep to the brain. As mentioned earlier in connection with fish, sleep is the absence of sensation. This occurs when the blood is cooled past the point of added sensation as illustrated below.

x hot=thick,passionate   x cool,thin=intellect   cold x
=thin, sleep

FIGURE FIVE: Temperature of Blood

When the heart's heat overpowers the blood, Man is passionate and excitable. The mean is considered rationality, while the brain's domination produces sleep. This model of fluctuating temperature resembles in many ways the Hippocratic theory of the four humours. In that account various illnesses could be explained by the predominence of one humour over another. It was imbalance which created problems so that the best possible state was a mean of proportioned balance.{82}

Aristotle seems to have borrowed much from the Hippocratics concerning the brain. The positioning of the brain in the front of the skull (656b 25) is found in On Wounds in the Head, c.2. Aristotle gives teleological reasons for this choice, but they seem particularly *ad hoc*. The brain is said to be in the forepart because sensation is there. But the brain has nothing to do with sensation, so this reason is weak.{83} Secondly, the brain's location is linked to the location of the heart in the front part of the body. But besides Aristotle's *a priori* disposition to the front over the back (as he showed in the right over the left) there is no mechanical reason why one's position need affect the other. His third reason concerns the sac of blood which he says is necessary for sensation. Because sensation occurs at the front of the head the blood sac must be there too. That is fine as far as it concerns the blood sac, but the brain's position has still het to be supported. Perhaps the only real reason offered is that the rear, empty cavity can act as an echo chamber for hearing.{84} I think, though, these reasons are contrived. Aristotle's actual reason for putting the brain in the front is because of the Hippocratic doctrine.{85}

Let us return to Aristotle's depiction of the brain as a cooling element. Aristotle believes that the brain must be composed of Earth and Water. This is because of the properties of Earth and Water.{86} Aristotle has already given a TE why the brain must be the cooling center of the body. If he can show that the brain is, in fact, composed of Earth and Water, then he can (on the basis of his physical theory) demonstrate via an ME that the brain is cold. Following Aristotle's usual method, he endeavors to offer both halves of the explanation. This demonstration takes the form of an experiment (653a 21).

1. Fruit and pulse when boiled become hard--observation.

2. Fruit and pulse are composed of Earth and Water, and when one boils them, the water is eliminated, leaving the Earth--from 1 & the definition of boiling.

3. The Brain when boiled becomes hard--observation.

4. The Brain is composed of Earth and Water--1, 2, by analogy to 3 & 4.

Aristotle moves from a supposedly known reaction to a remote one. When one notices the same effect upon the remote thing that one noticed from the common item, then one is convinced by hypothetical syllogism that the unknown quantity also has said property. This method is sound,{87} but as before, the application of the method is specious. Fruit and pulse when boiled do not become hard, they become soft. Perhaps Aristotle is referring to some sort of baking (though epsēma does refer to boiling, ll. 22, 24). It is difficult to know what Aristotle wants to show by this proof. Could it be that baking fruit and a brain until each are charred really show a similarity? Would not any food act in the same way? Juicy fruit is appropriate as an analogy if Aristotle wants to claim that the brain is cool and moist, but while the principle behind his experiment seems to be soundly based, his implementation of this experiment is so vague (and seemingly incorrect as worked) that it is difficult to arrive at a specific conclusion from it.

Throughout this section on the uniform parts Aristotle's arguments have been shown to be characteristically less tied to careful repeated observations than was evident in his accounts of bloodless animals (particularly marine animals). It is not the case that Aristotle has altered his method, rather it is the case that in carrying out this method Aristotle seems rather in a hurry to come to conclusions by not proceeding as carefully as he should have (i.e., the way he did with the bloodless animals). Perhaps the human body interested Aristotle less as a biological problem and more as a series of philosophical problems concerning the relationship between body and soul via De anima.{88} Whatever the case may be, these lapses are a principal reason for some of the less satisfactory conclusions arrived at in his description of the uniform parts.

4.7 NON-UNIFORM PARTS

The non-uniform parts include external and internal parts. Some external parts are: parts of the face, the face itself, the arm, hand, horns and the like. Internal parts are principally the various organs: heart, liver, spleen, lungs etc. As mentioned earlier the difference between the uniform and non-uniform parts lies in their organization. A non-uniform part has a heterogeneous organization. Any part with a specialized function is likely to be non-uniform. A particular bone, like the tibia, is non-uniform, while bone itself is uniform. Thus one sees that the non-uniform parts, by definition, are naturally no different than the uniform parts. It is the uniform parts which make up the non-uniform parts. What is different is the organization and particular configuration which is adopted especially for one or a limited number of functions. One example of each type of part will be examined to conclude this discussion of Aristotle's practice in his biological works.

Horns are external non-uniform parts. They are found only on vivipara (662b 25). Even when other animals have appendages that resemble horns, these cannot be classified as horns since their function is not for self-defense or attack. Non-uniform parts are defined in terms of their ergon, and Aristotle is consistent with his broad definition here.

Nature gives animals a means for defense but never more than one (663a 16). Therefore, horses and elephants do not have horns because their speed and size, respectively, afford them ample means of protection. But to this general rule Aristotle offers some exceptions. Reading from Herodotus (4, 183) Aristotle cites the story about the backward-grazing oxen (659a 21). These animals' horns are large and make it impossible for the oxen to walk forward lest their horns stick into the ground. Therefore the oxen walk backwards. These animals seem to be encumbered rather than aided by their horns, which is contrary to Aristotle's teleology.{89} The deer's horns also are useless for self-defense or for fighting. Deer have speed to escape predators. Aristotle correctly notices that because of the thin, branch-like configuration of their antlers, the deer's horns are not good for actual combat (since when one buck fights another their horns often become entangled requiring the fighters some trouble to extricate themselves). There is a usefulness in the deer's horns for "bluff" fighting where one makes a show to fight and the other backs down (though Aristotle does not point this out). Just as in birds, this bluff fighting constitutes a mating ritual, but such behavior was not noticed by Aristotle. He saw the horns of the deer and backward grazing oxen (along with the talons of certain birds 694a 20). How does Aristotle account for these exceptions? It is not clear. Aristotle states that "to study nature we have to consider the majority of cases, for it is either in what is universal or what happens in the majority of cases that Nature's ways are to be found" (663b 17). Aristotle is trying to come to grips with physical laws. These are not numerous (if to be found at all) in the biological sphere. Just as Aristotle admitted that the rinobatus might be an exception to fish mating (4.3), so here Aristotle is willing to admit that exceptions might exist. The status of his "laws" seem more on the order of statistical generalizations which accept exceptions in their very format. Two factors must be considered. First there is the process by which one acquires <u>epistēmē</u>, as outlined in the <u>Posterior Analytics</u>. The knowledge of the conclusion of a demonstration yields epistēmē. It is clear that one may have <u>epistēmē</u> about the causes of "normal" events (i.e., events not interrupted by force or violence). For Aristotle's biological works are in accord with his philosophy of science in a general way. But secondly, there can be considerable difficulty in understanding the "<u>hoti</u>" when it comes to biological examples. This makes many laws impossible on the grounds stated in APo. 1, 2 & 3. In these cases, one, as a biologist, considers the majority of cases, so that some kind of generalization can be generated. This is to say that the rule is not absolute (since the <u>hoti</u> is not exact, neither can the <u>dioti</u>, cf. APo. 2.1 and Aristotle's example of lunar eclipses) and that it holds only for the most part.

The reason the <u>hoti</u> cannot always be known is that nature/Nature is often so complicated that rules can have only partial dominion. Because there is sufficient variation within the animal realm, full "laws" in the sense of APo. are generally impossible. But this is a practical concern. In principle, any form should be comprehensible according to some conditionally determined process. In Aristotle's time there were many gaps in practical biological understanding (even as there are today). These views are important aspects of scientific recognition necessary for sophisticated biological study. Aristotle may have had his "lapses," but such judgments should not overshadow his overall biological achievement.

Aristotle also observed a connection between the nature of the hoof and the number of horns. Cloven-footed animals tend to have two horns, while solid-hooved animals tend to have one horn (the exception to this rule being the Oryx and the "Indian Ass," 663a 20-5). Aristotle's discussion of the single-horned animal takes two forms.

> A. Now it is quite reasonable that the one horn should go with the solid hoof rather than with the cloven hoof, because hoof is identical in nature with horn, and we

> should expect to find divided hoofs and divided horns together in the same animal. B. Again, division of the hoof is really due to deficiency of material, so it is reasonable that as Nature has used more material in the hoofs of the solid-hoofed animals, she has taken something away from the upper parts and made one horn only. (P.A. 663a 27-33).

The first explanation is teleological, from the point of view of Nature as an artist. Nature strives after bilateral symmetry. Putting the horn in the middle serves this purpose as best as the material will allow (663a 25). The second account has two parts which describe in increasingly mechanical terms how necessity interacts with teleology in the formation of one horn. Given that the animal has one horn it must be the case that there was a deficiency of material available, so that this deficiency was divided between the horns and the hooves providing equal constraints on both. In this case neither Aristotle's TE or ME is sufficient alone to describe the phenomena, so both act together to form the explanans.

This dual form of explanation is repeated when Aristotle discusses why the horns could not have been placed on the shoulder as playfully suggested in Aesop's fable. If the purpose of horns is to fight, then horns on the head offer greater mobility in aiming and directing the blow. (663b). Mechanically, the position of horns on any other appendage would create a weight distribution problem which would also hinder the animal in carrying out its purpose.

In the same way after Aristotle has given his TE on why only the deer's horns are shed (663b 15 viz., for mobility as their natural form of defense is speed), he identifies his shift from TE to ME:

> This completes our statement of the purpose for which horns exist and the reason why some animals have them and some have not. We must now describe the character of that "necessary nature," owing to which certain things are present of necessity, things which have been used by "rational Nature" to subserve a "purpose." (663b 222-24).

Aristotle goes on to offer an account of why some animals have horns through a description of the movement of surplus corporeal matter (ME).

THE PRACTICE

Aristotle's description and account of horns offers an excellent example of TE and ME acting together to explain what one might think would be peculiarly an area for TE, since non-uniform parts are defined with respect to function. Aristotle seems to think both are necessary—an attitude which carries into his discussion of the heart. Here the principle of bilateral symmetry is stretched slightly because of the phenomena. To begin there is one heart and it is located off center. Though Aristotle does link the heart and brain as reciprocal organs, this does not satisfy the bilateral symmetry that he achieves by matching the liver and spleen (669b 27).{90}

The actual construction of the heart into three chambers seems to have been motivated by observation rather than by a yearning toward symmetry even though Aristotle tries to make three chambers consistent with the principle of symmetry.

> Now there are two chief blood-vessels, the so-called Great Blood vessel, and the Aorta. Each of these is the source of other blood vessels. The two differ from each other (this will be discussed later); hence, it is better for them to have separate sources. This result can be obtained by having two separate supplies of blood, and thus we find two receptacles wherever this is possible, as in the larger animals which of course have large hearts. But it is better still to have three cavities, and then there is an odd one in the middle which can be a common source for the other two; since however, this requires the heart to be particularly large, only the very largest hearts have three cavities. (666b 26-35).

Aristotle's reason for the middle chamber on the grounds of the right/left distinction seems contrived especially when one remembers his discussion of 'oddness' regarding the sea urchin's ova. An even number would have worked out better for Aristotle, but Aristotle apparently held his conviction on the basis of actual dissections.{91} Now, dissecting a dead animal, especially one killed by strangulation (recommended by Aristotle at H.A. 513a 13){92}, can create problems in viewing all the chambers properly. This is because the superior vena cava and the inferior vena cava collapse, causing the right auricle to also collapse.{93} This would lead Aristotle (and even Galen) to suppose that the right auricle was only a junction of the two vena cavae and not a part of the heart. By such an account one might expect the heart to be pictured as in figure 5-b. This is the position taken by most modern commentators.{74} The key passage comes from H.A. 513a 27:

The heart in all animals has cavities inside it. In the case of the smaller animals even the largest of the chambers is scarcely discernible; the second larger is scarcely discernible in animals of medium size; but in the largest animals all three chambers are distinctly seen. In the heart then (with its pointed end directed frontwards, as has been observed) the largest of the three chambers is on the right-hand side and highest up; the least one is on the left-hand side; and the medium sized one lies in betwixt the other two; the largest one of the three chambers is a great deal larger than either of the two others. All three, however, are connected with passages leading in the direction of the lung, but all these communications are indistinctly discernible by reason of their minuteness, except one. The great blood vessel then is attached to the biggest of the three chambers, the one that lies uppermost and on the right-hand side; it then extends right through the chamber, coming out as blood vessel again; just as though the cavity of the heart were a part of the vessel, in which the blood broadens its channel as a river that widens out in a lake. The aorta is attached to the middle chamber; only, by the way, it is connected with it by a much narrower pipe.

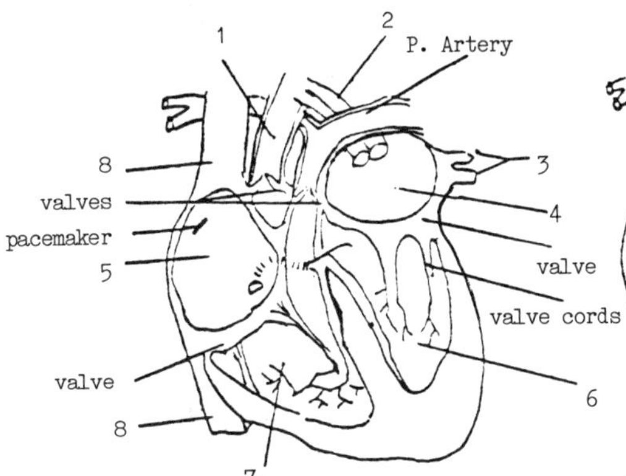

a) The heart as pictured today

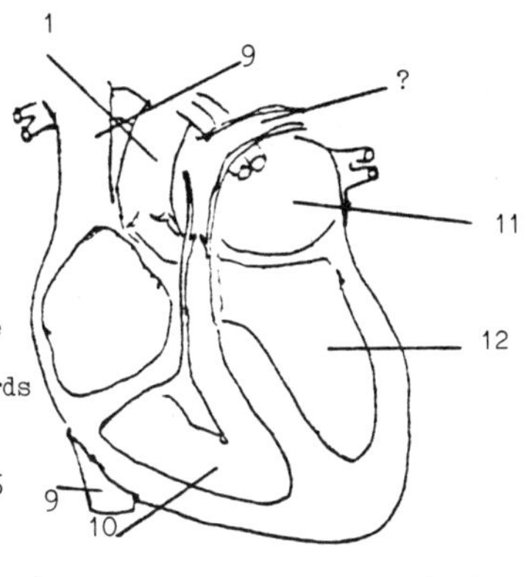

b) Reconstruction of the heart via Lones, Ogle, Huxley; and Platt

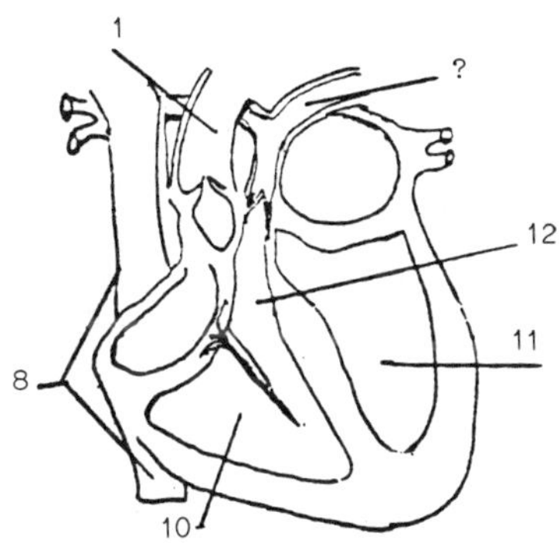

c) Galen's version on a heart with partially collapsed right auricle

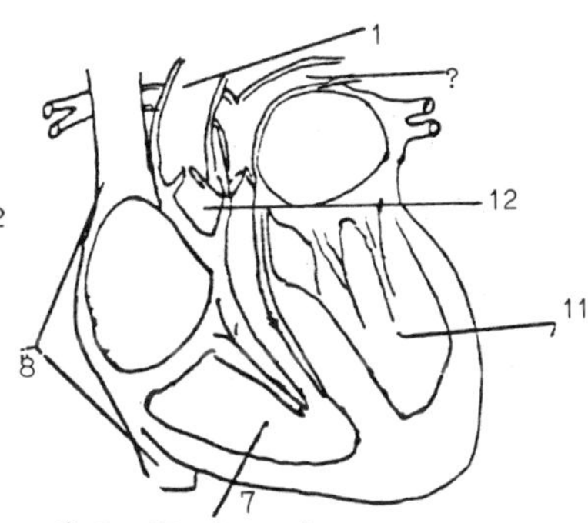

d) Vesalius' version

1. Aorta
2. Right Pulmonary Veins
3. Left Pulmonary Veins
4. Left Auricle
5. Right Auricle
6. Left Ventricle
7. Right Ventricle
8. Vena Cavae
9. Great Vessel
10. Right Chamber
11. Left Chamber
12. Middle Chamber

FIGURE SIX: Views of the Heart

Diagram (b) fits Aristotle's statement that the right chamber is the largest and that the left is the smallest. Also, it allows the aorta to depart from the middle chamber. The problem with this account is that Aristotle only mentions two major blood vessels entering or exiting the heart, megalēn kaloumenēn kai tēn aortēn (P.A. 668b 26, cf. H.A. 513b 5). The so-called great blood vessel is probably the vena cavae as they connect in the right auricle. This is evidenced by the description of it lying in front of the aorta, P.A. 668a 2. If this reading is correct, then there is no provision made for the pulmonary artery which sticks out of the heart as a major vessel. It is this fact that led Galen to picture Aristotle's middle part of the heart as a part of the right ventricle: "the third cavity is that at the broad end of the heart, being part of the right chamber and not a distinct third chamber."{95} From this reading the middle cavity is directly between the two ventricles on a horizontal axis. Aristotle's statements about the middle being a common source to the right and left side might be thus explainable (666b 32, cf. De somno ch. 3).{96} Vesalius offered a similar reading except that the middle cavity was a part of the left ventricle instead.{97}

Galen's interpretation does not agree with Aristotle's statement that the aorta passes from the middle chamber and so may be discounted as an accurate interpretation of this passage from the Historia animalium. Vesalius' account where the aortic vestibule is the middle chamber agrees with the passage about the aorta's attachment, but this reading does not account for Aristotle's statement that the left chamber is the smallest, for Vesalius' reading (as well as Galen's) makes the middle chamber to be the smallest.

Thus the first reconstruction seems to be the most reasonable. But there is still the problem of the pulmonary artery. Two solutions suggest themselves. First Aristotle does mention that all three chambers have passage ways connecting to the heart which are very small save one. That one could be the pulmonary artery. If this is true, it is strange that Aristotle did not grant the pulmonary artery greater attention since it is not appreciably smaller than the aorta. Now it could be argued that the pulmonary artery is simply a continuation of the vena cava. The passage at 513b 7, "The great blood vessel passes through the heart," seems to suggest this interpretation. {98} However, as Thompson points out (note 513a 35), this is problematic because it supposes that Aristotle "was aware that the venous circulation of the body was continued directly through the heart into the pulmonary circulation." Also it seems odd that Aristotle would have thought an artery was a vein in view of the fact that he describes their physical difference so accurately at 513b 8. Thompson suggests that both the pulmonary artery and aorta were viewed together as the "aorta," i.e., part of the arterial system leading from the heart. There are merits to this proposal, however; in the discussion of blood vessels in the _De partibus animalium_ Aristotle makes it clear that he believes the crucial importance of these two main vessels is that they lead to the rest of the body (668b 20ff.). Aristotle goes on to outline the way in which the whole body is serviced by branches of these two major vessels. Now the pulmonary artery goes to the lung and therefore does not fit into the scheme Aristotle seems to have in mind for the arteries. {99} What seems likely to me is that the pulmonary artery is merely the connection between the heart and lung described at H.A. 573a 35. The size of the pulmonary artery might be accounted for by the fact that it is servicing the largest chamber (right ventricle). The smaller passages described are perhaps for _pneuma_ or for cooling ducts operating directly from the lungs.

The heart is described by Aristotle as the _archē_ of the body. It holds a special place among all parts as it is both a uniform and a non-uniform part (647a 25, cf. _De somno_ 455b 34). There are, according to Aristotelian methodology, two ways in which something may be said to be an _archē_: 1. It can be first in _logos_ and 2. It can be first in development (for the discussion of these points see 3.2). Aristotle wants to claim that the heart is prior in both ways. To prove the first point he must return to his definition of an animal. An animal is something that has the power of sensation (P.A. 653b 22). Aristotle does not tell the reader how he arrives at this definition except to say that he constructed a hierarchy of living things and the lowest, plants, seem only to absorb food while animals seem to have another capacity.

> This then, is the principle through which all living things have life, but the first characteristic of an animal is sensation; for even those which do not move or change their place, but have sensation, we call living creatures, and do not merely say that they live (_De anima_ 413b 1-4).

Sensation, therefore, is an added power which animals have and plants do not.{100} This conclusion seems to be made from examining plants and animals and trying to discern the various functions each performs. The power of sensation is found in heat:

> But fire is the medium of no perception, or else is common to them all (for there is no possibility of perception without heat) . . . . (*De anima* 425a 5-6).

The reason Aristotle chooses heat may be due to the fact that heat has the power to change food into something edible, and from this grows Aristotle's analogy of digestion to cooking (*De anima* 416b 30, *De sensu* 442a 5). Heat can "concoct" food and bring forth nourishment. This transforming power seems to make heat more active than cold. When a pan is heated, lard crackles and motion occurs.{101} Heat, therefore, would appear to be the proper element to be responsible for motion. Secondly, a major physiological difference between a corpse and a live body is the loss of heat. If loss of heat is involved in death, and death is the loss of sensation, then sensation must be caused by heat (*Youth, Old Age, Life and Death* 469b 21).{102}

This last argument depends upon the unstated premise that there is one and only one variable connected with death so that any external changes in the corpse must be attributable to that one variable, such as dilated pupils etc. If there are several factors arranged equally or in a hierarchy, then this argument does not work. For example, if heat were the result of the respiratory cycle and connected with the heart only as an interdependent system and if sensation results from nerves or *pneuma*, then when the body dies heat will stop, the body will grow cold and sensation will cease. All happens concurrently, but this nexus of disfunction does not imply that all these factors are the same or that any one of them is a simple cause for the loss of sensation. The heart and respiratory system are interdependent, but this does not allow the direct association of heat, sensation and the heart. The error involved here is a logical one in the interpretation of the data.

Given that Aristotle does make this association it is easy to see how the heart becomes the principle of sensation since it is identified as the primary source of heat in the body.{103}

> Indeed, the heat that is in the heart, being the source, is extremely responsive to any influence upon it . . . the correct view is that the seat and source of sensation is in the region of the heart . . . an animal is defined by the fact that it possesses sensation: and the part of the body to have sensation first is the part that has blood in it first—in other words, the heart, which is the source of the blood and the first part to have it. (653b5, 656a 28, 666a 35-666b 1).

As the primary source for heat, which is the cause of sensation, that soul which makes animals alive, the heart, is the principle of life itself.{104}

This first aspect of the heart's primacy has depended upon a combination of observations and rational inferences. Aristotle's observations are simple and direct, but some of his conclusions, which he draws from his data, are not justified. Still, the overall conclusion seems correct despite difficulties in his argument to support it (is this only luck, given his faulty argument?).

The second sense in which the heart can be the <u>archē</u> of the body is in genesis. Aristotle wants to show that the heart is the first organ formed and thus prior in another sense. Aristotle means to prove this by examining the eggs of a common chicken and cracking open one egg a day to observe the developing embryo. He reports his results in several places:

> The first to be formed is the "principle," which in blooded animals is the heart and in others the counterpart of the heart, as I have said many times over. There can be no doubt about this, because our senses tell us that it is the first thing formed; but the truth of it is confirmed by what happens when the creature dies: the heart is the place where life fails last of all; and we find universally that what is the last to be formed is the first to fail, and the first to be formed is the last to fail . . . . . In embryos, as soon as they are formed, the heart can be seen moving before any of the other parts, just like a living creature . . . the heart and the liver are visible in blooded animals as soon as they are formed at all, that is, when they are quite small: in eggs they are visible, just about the size of a point, sometimes as early as the third day, and very small ones are visible in aborted embryos . . . . With the common hen after three days and three nights there is the first indication of the embryo . . . where the primal element of the egg is situated . . . the heart appears, like a speck of blood in the white of the egg. (G.A. 741b 15ff., P.A. 666a 2—22, 665a 33-35, H.A. 561a 5, 9, 11).

Aristotle's question is, "What organ really does come first in genesis?" The question is strictly empirical and deserves an investigation of actual embryonic development. This Aristotle does, noticing that on the third day a red spot appears showing him the heart is formed first.{105} Had Aristotle been acquainted with staining techniques he would have found that the spine and nervous system develop first, but with his methods and materials, his observation is correct. Aristotle's work here seems to be original, and this alone entitles him to a seat among the great pioneers of biology. He uses these observations not only to prove that the heart is formed first, but also to show that the embryo was not preformed, but rather each part was developed in hierarchical fashion.{106}

It is interesting to note that in the passage from the *De generatione animalium*, Aristotle does not feel that his original observation was enough, so he adds a second argument that the heart is the last organ to die. Since "the last to be formed is the last to fail" and the heart is the last to fail, therefore, the heart must be the first formed. This argument follows Aristotle's inclination to mix empirical and rational arguments together just as he mixes ME and TE. No support is given for his maxim which must be a common opinion. No evidence is given that the heart is the last to fail. In fact, if death is defined by the activity of the heart as it seems to be above, then it is after the heart stops that all else stops. When the heart fails, then all else fails, i.e., the heart is the first organ or (in cases of disease or accident) one of the first organs to fail. Aristotle's second argument is very weak, but it should not overshadow the genuine achievement of his embryonic studies of chicken eggs.

Thus Aristotle offers two arguments that the heart is the *archē* of the body. Each is unique, though they both employ techniques of investigation which we have encountered earlier. Aristotle's work on the heart is very interesting. He seems to have had first-hand knowledge of several animal hearts about which his accounts are most accurate. His methodology is consistent even when he is discussing humans, though like his work on the brain, his human physiology seems less careful than his animal physiology.

Notes on Chapter Four

1 This can include common systems as well, e.g., the digestive system.

2 My translations from the *Historia animalium* will follow D'Arcy Thompson's rather closely; D'Arcy Thompson tr. *The Works of Aristotle* vol. 7, (Oxford, 1910).

3 Aristotle neither notices the simple eyes of insects nor the nature of the compound eyes.

4 This seems particularly true from his descriptions of the internal organs.

5 Cf. P.A. 682b 20. The segments here are presented as "for the sake of" an insect's curling up.

6 The distinction is now termed "coiled" vrs. "uncoiled." There are also two guts (fore and hind), a crop, gullet, gizzard and tubular stomach.

7 Hearing in some insects is located in the abdomen.

8 Aristotlo might say, "What can taste be for except eating?" Eating is done via the mouth, therefore; taste must be associated with the mouth. If the sting brings food to the mouth, then the sting must be associated with taste. But what if taste and smell are used as a method of finding before food is ever obtained. In this case, these possible justifications would not be necessary.

9
I have termed this observation "regeneration" because is the manner of which most insects respond to mutilation (if they successfully respond at all). Three possibilities can occur after mutilation: a) the insect can continue without said part(s)--if it is fortunate enough to have multiple legs, for instance, there may be no loss of functionality if one leg is removed; b) the insect may regenerate said part; c) the insect may die. From the language it is unclear whether he means simply (a) or (a) & (b). Either way, the reason for continued existence is due to the _archē_ argument.

10
Of course Aristotle mistakes the crucial facts of the heart being the cause of this phenomenon (and not the stomach). He does say, though, that the stomach is the analogue of the heart for the bloodless animals. This mistake is understandable since Aristotle had no idea of the real function of the heart which would enable him to compare it to other animals, such as the insect. The color of the blood is a critical reason for this. Seeing no red blood Aristotle had no choice but to deny the existence of a heart. Also, Aristotle misses the point of decreased cell specialization. On the point of division of labor see Ogle's note: 683a.

11
See: Peck's note on imperfect eggs from his introduction to the _Generation of Animals_, # 77, p. lxx.

12
The wool is formed from the dust on clothes (a very good reason to keep one's clothes tidy!). G.A. 758b 22, H.A. 557b 2.

13
The most easily accessible insect that I have observed copulating is the dragon fly, which, because of its size and choice of location (the banks of lakes) is not difficult for the naturalist to observe.

14
I have discussed this process in greater detail in "Aristotle's Digestive and 'Circulatory' System" _Journal of the History of Biology_ 15.1 (1982): 89-118.

15
This connection is not as simple as Aristotle makes out. He should have seen that from the data he had he is not justified in drawing the conclusion that he did; cf. the discussion of the heart in 4.7 and Boylan (1982).

16
Pliny, 11.22.

17
There is some discussion on this passage: _di hēdonēn touto poiousin he dia phobon_. See Dittmeyer and Thompson on this.

18
The Queen is a fully developed female. The worker is a partially developed female and the drone is a male. The same eggs produce workers and queens, except that the latter is treated with royal jelly, a food produced by the workers. The drone eggs are unfertilized eggs in contrast to the female eggs. See Peck's note: 759a.

19
These terms have been used to translate Aristotle's terms here in accordance with Peck and the standard usage: _ta ostrakoderma_ = testacea; _ta malakostraka_ = crustacea; _ta malakia_ = cephalopods. See also _Chapter Two_ and the chart _of the genera_.

20
There is an extended discussion on the history of this discovery by Charles Singer, "Greek Biology and its Relation to the Rise of Modern Biology" _Studies in the History and Method of Science_ ed. Charles Singer (Oxford, 1921), pp. 41-46.

21
See Thompson's note 524a, and his diagram, 524a 4.

22
See Peck's note 524a.

23
The fact that beekeepers' observations are pragmatically linked to honey production etc., may have influenced their reports [and what they actually were looking for]. This fact did not seem to bother Aristotle, though perhaps it should have.

24
Cf. maxim on generality 571a 21; 4.5.

25
This is not always the case, as Aristotle states in his example of octopus fear, which is described as simply being a direct sign stimulus and fixed action patter response. The octopus sees something and becomes frightened discharging the residue from its bladder, P.A. 679a 25.

26
For this point I am assuming that the _Historia animalium_ is prior to the _Generatione animalium_. This is argued by D'Arcy Thompson in the introduction to his translation of the _Historia_ and by H.D.P. Lee, "Place-names and the Date of Aristotle's Biological Works" _Classical Quarterly_ 49 (1948): 61-67. An opposing view is being established presently in a yet unpublished work by David Balme. I have put the works into the following order: H.A., P.A., G.A., De.An. (the order of the other works is not important to my argument).

27
Not until recently has this process been discovered. The spermatophore when exposed to seawater, suddenly releases a spirally wound, springlike mechanism which begins to uncoil, stretching out the tubular portion of the spermatophore to a much greater length. The sperm become active and come boiling up the tube as if under pressure. Thus, Aristotle (or anyone else for sometime) could not determine the existence of this passage because it happens only upon the events directly preceding copulation. See: Bayard H. McConnaughey, _Introduction to Marine Biology_ (St. Louis, Mo., 1974), p. 188.

28
Aristotle could have examined an arm that had been detached upon the female for evidence of a spermatophore. This would have given him conclusive evidence on the question. This was, in fact, the way the event was rediscovered (see: Singer, p.44).

29
Peck discusses this in his introduction to the _Historia_, p. 13.

30
H.A. 524b 31, cf. Thompson's note.

31
Thompson discusses this in The Legacy of Greece (Oxford, 1921) in his chapter "Aristotle the Naturalist." Apparently there was some confusion about this passage which was due to the fact that directions "right" and "left" apply to a lobster which was laid on its back.

32
One might try to explain this away by appealing to the inherent properties matter has. This could be merely an instance in which the matter was not "mastered." However, this tack cannot be accepted because there is no device to insure the regularity of formation present in lobster claws.

33
One interesting observation that Aristotle made at H.A. 531a, cf. P.A. 680a 5, concerns the famous "lantern." This lantern is supposedly the mouth apparatus which looks like an ancient lantern. There has been some literature which has grown up around this point. See: F.J. Cole, Centaurus 1 (1950): 377; T.E. Lones, Aristotle's Researches in Natural Science (London, 1912), p. 127, 130; A.L. Peck, "Additional Note on Sea Urchin's Lantern" from his translation of the Historia animalium, p. 351.

34
There is no connection here between teeth, stomachs and ova. It would seem that Aristotle would agree to this point in light of his criticism of Herodotus at G.A. 756b 5. One wonders what attraction Aristotle has in this context with the number five.

35
There is a minor error in his mistaking a stomach for a bladder (Thompson).

36
Cf. Platt's note and Peck's note to G.A. 746b 5. The rhine is called an "angel fish" by Thompson and a "shark" by Platt.

37
Aristotle does not discuss the status of such a rule nor the relative confidence that one can have in a rule that is only true "for the most part." Is there any difference between such a generalization and one which says, "All the bears in the North

Pole are white," or "All the coins in my pocket on January 1, 1983 are silver." The point is to determine how far Aristotle wants to press his idea of a rule (law?). For a further discussion on these issues see: Rudolf Carnap, Meaning and Necessity: A Study in Semantica and Modal Logic (Chicago, 1947).

38
There are some textual problems here which are reproduced by Peck on his note on 537a 3. Dittmeyer's text seems to be the most reasonable and was adopted for my reading.

39
There is the question of the sensitive 'soul,' but the character of this is not connected with sleep (avoiding circularity), nor is it invoked to verify soul. 'Soul' merely explains the limits of our observations.

40
Cf. Peck's note.

41
Another interesting distinction that Aristotle made concerns internal eggs. Of course Aristotle knew nothing of mammalian eggs, which are singly only visible under a microscope. Aristotle also speaks of an internally oviparous and externally viviparous organism (fish) at 754a 24.

42
I am not talking about the 'selfish gene' hypothesis as put forth by sociobiologists such as Richard Dawkins, The Selfish Gene (N.Y.,1976) or E.O. Wilson's, Sociobiology: The New Synthesis (Cambridge, Mass., 1975). Rather, the point here is merely to posit a higher level of organization in accordance with the observed group characteristics.

43
At G.A-. 755a 18 Aristotle also gives an ME why the eggs grow.

44
Peck calls these "dualizers." Cf. my note 67 and Preus (1975), pp. 159-162.

45 Aristotle breaks up birds into non-fliers, fliers (both large and small), and fliers with crooked talons (G.A. 749b 1-25).

46 Cf. Peck's note in his appendix to G.A. p. 588. The casing of an egg which is 'perfect' cuts it off from the environment so that all necessary ingredients must be included within the shell. Recent studies concerning the relationship between eggs and their environment have indicated that bird embryos take up oxygen and discharge carbon dioxide by simple diffusion through microscopic pores in the eggshell. The process is regulated largely by pore geometry. See: H. Rahn, C.V. Paganelli and A. Ar, "The Avian Egg: Air-Cell Gas Tension, Metabolism and Incubation Time" _Respiration Physiology_ 22 (1974): 297-309; "How Bird Eggs Breathe" _Scientific American_ 240.2 (1979): 46-56.

47 Peck thinks this can be explained via _Mech._ 848a. I discuss this whole process in more detail in my article "The Hippocratic and Galenic Challenges to Aristotle's Conception Theory" _Journal of the History of Biology_ (1983)--forthcoming.

48 Cf. Platt's note on the Hippocratic debate on this issue.

49 It was William Harvey who demonstrated that the cicatricula was the point of origin of the embryo.

50 Platt partially confirms Aristotle's observation by citing the composition of duck eggs which he says demonstrates this.

51 D'Arcy Thompson, _A Glossary of Greek Birds_ (London, 1895).

52 Many birds cited uniquely by Aristotle still indicate to Thompson a foreign origin in mythology.

53 Cf. Thompson, p. 46.

54
For further references see Thompson: pp. 8, 28, 31, 63, 107, 121, 132, and 192.

55
One should contrast what Thompson is describing here to legitimate connections which have been shown to exist between birds and the stars. For example, in Matthews' famous experiments with bird migration he used a planetarium to change the star patterns which the birds could see. By simply altering the star patterns, he could induce the migratory behavior in weeks (complete with added fat deposits, enlarged gonads and restlessness). This latter process takes far longer. He could also induce the birds to position themselves in their cages (on the perches) according to the star patterns. Geoffry Vernon Townsend Matthews, Bird Navigation, 2nd ed. (Cambridge, 1968). This position is different from unsubstantiated claims which have only astrology or mythology as their basis.

56
In Alan Gewirth's, "Positive 'Ethics' and Normative 'Science'" Philosophical Review 69 (1968) the point is made that distinctions are made in science between what will be accepted as "scientific" and what will not. One does not view all data equally.

57
Among those who have rejected the ninth book are Aubert and Wimmer who claim that at best it was put together by an editor from disorderly notes left by Aristotle. H. Joachim, De Theophrasti libris PERI ZOON (Bonn, 1892) calls it a later work by a peripatetic after Theophrastus. Dittmeyer, Aristotelis De animalibus historia (Lipsiae, 1907) incorporates aspects of both of these arguments. On the other side fall J. Tricot translator of the Bude edition (1957) who follows Thompson in cosidering the ninth book to be authentic.

58
In other words the pharynx and lungs are parts which contribute to the ME of the voice.

59
"The Indian bird, the parrot, which is said to have a man's tongue, answers to this description; and, by the way, after drinking wine, the parrot becomes saucier than ever." (H.A. 587b 27).

60
W.H. Thorpe, Bird Song: The Biology of Vocal Communication and Expression in Birds (Cambridge, 1961), ch. 4 & 5.

61
ibid., p. 73.

62
Cf. Thompson's note. One suggestion is that the creature in question is a sea slug, though I rather doubt it.

63
The height a projectile will go is proportional to the square of its velocity at the surface. Thus, unless the boats have extremely low masts or the dolphins move at very high velocities, this can be no more than a "fish story."

64
See Preus (1975), n.43, p. 268; p. 23.

65
Cf. Preus (1975), n. 78, p. 310.

66
One modern analogue to the problem between plants and animals can be found in the Euglena. Because it has chlorophyll it seems like a plant, but it can survive without photosynthesis. In fact it may divide so rapidly under such circumstances that some offspring fail to get a chloroplast. It has a flagellum so that it is capable of locomotion. The only way that one can deal with such cases in evolutionary systematics is to adopt one trait as crucial. Otherwise a "neat" classification cannot be made (a fact that would not have disturbed Aristotle, who was willing to alter his systematics to fit the phenomena).

67
Intermediates between animals and plants include Ascidians, Sponges, Holothuria, and Acalephae. Bats and Ostrich are between quadrupeds and birds. See also H.A. Peck's note on "dualizers" in his introduction to the Historia, p. lxxiii.

68
This is contrary to the Hippocratic doctrine that pneuma itself makes the skin sensitive by being given phronēsis. This account seems to have been toyed with in the earlier passage from P.A., but rejected in the later passage from De anima. Of course a common sense is needed, but the point is more

complicated than this. What is it that "feels"? Granted the sensitive soul registers this feeling, but is there any feeling <u>in</u> the finger which is pricked by a pin? What is the throbbing afterwards if there is not something that is happening <u>in</u> the finger. Aristotle never comes to grips fully with this question.

69

"In neither class of the cases mentioned would sensation result from touching the sense organ; for instance if one were to put a white body on the surface of the eye. From this it is clear that that which is perceptive of what is touched is within. This is true in the other cases; for when objects are placed on the sense organs no sensation occurs, but when they are placed on flesh it does" (*De anima* 423b 20-26). This is an empirical example to prove Aristotle's point here.

70

Ichor and *hines* appear at *Timeaus* 83c 5 and 85c 3, d 2 respectively. In this context, "serum" is identified as the rare innocent part of the blood, whereas "fibers" are said to 1) maintain the balance between the rare and the dense and 2) congeal the blood. Obviously Aristotle picked up on the "rareness" of 'serum,' but concerning 'fibers' he ha adopted the second usage and ignored the first. In fact, he has made the 'fibers' to be the dense part when Plato portrayed this function by phlegm. Another interesting cross reference was suggested to me by Arthur Adkins. In *Iliad* 5.340, 416 *ichor* is used to describe the blood of the gods. It is perhaps here that the noble, pure connotation for *ichor* was derived. Though this parallel is interesting, there are also some difficulties, e.g., the temperature of the two vary. M.P. Duminil has made an interesting case for the evolution of the term from Homer through two Hippocratic systems (the 'noxious fluid' system, and the humours) to the use of the word as a fluid element. "Les Sens de *ichor* dans les Textes Hippocratic" *Corpus Hippocraticum* (Mons, 1977), pp. 65-76.

71

This seems a little like the barley gruel example cited in Chapter One. That is, material properties are being ascribed to explain something mechanically, though no justification is given.

72
Likewise the front is over the back and the upper over the lower. For the right over the left see: 647a 10. On the front over the back see the growth patterns of hair at 658a 23, and the heart's position justified 665b 18. That the upper is preferred over the lower is shown at 665b 20 and 686a 13.

73
Preus suggests that the origin of this is from Pythagorean tables, p. 219.

74
In De anima 405b 28 Empedocles uses a priori principles of natural direction to explain growth of plants and is criticized. But Aristotle is doing the same thing in G.A. 766b 35 with his theory of right over left to explain semen concoction. This indicates perhaps that this a priori direction is unintentional as it is contrary to his method and his application of it as applied to others.

75
This Hippocratic doctrine might also originate from the sources suggested in note 70.

76
Ogle quotes Thackrah in his note to this passage from (On The Blood, p. 154) that "from my observations the general inference may be drawn that coagulation commences sooner in small and weak animals than in large and strong." This is in direct opposition to Aristotle's statement.

77
Lones argues this convincingly in his chapter on dissections.

78
Lones, p. 179, suggests that Aristotle only saw a cooked brain.

79
As far as visual analysis allows, this is still unsatisfactory. How Aristotle could have asserted that the optic nerve (clearly visible on any large animal) was not "in contact" with the brain is difficult to understand.

80
It is interesting to note that the brain does act as a cooling mechanism in certain animals such as the cat, sheep, dog and certain varieties of monkey. This is accomplished by the blood traveling through the carotid _rete_ before going through the circle of Willis. It is within this network that the warmer arterial blood loses heat to the cooler venous blood. It is interesting to note that this basic structure was known (anatomically, though not physiologically) as early as Galen and possibly Herophilus. For a discussion of this phenomenon see M.A. Baker, "Influence of the Carotid Rete on Brain Temperatures in Cats Exposed to Hot Environments" _Journal of Physiology_ 220.3 (1972): 711-728. Preus makes an interesting comment on the possibility of there being a uniform correct temperature for the body. If such a temperature existed there would be no need for a cooling mechanism. However, outside influences such as climate and food ingestion (Preus, 1975, p. 122), affect temperature (cf., _Airs, Waters and Places_ discussed in Chapter One).

81
The stomach is called the third major organ and is paired with the heart and the brain (_On Youth and Old Age, On Life and Death_ 69a ff.). This is perhaps because the brain, heart and stomach all fall in the middle of the body and break the natural bilateral symmetry that is so attractive to Aristotle.

82
Cf. On Ancient Medicine and the discussion of balance between the humours.

83
Cf. notes 70 and 75.

84
P.A. 656b 14; De.An. 419b 32, 425a 4.

85
Because of the earth-water mixture the brain is moist and cool. The properties come from the elements involved. This is also given as a reason for the slow hardening of the bregma at P.A. 653a 30. Since heat is necessary for bone to harden and since the brain is cool, this bone takes the longest to harden.

86
A discussion of the interaction of the physical properties of the contraries can be found in G.C. 330a 30ff.

87
Except in the case where one effect has multiple causes, e.g., (A v B) => C; D => C; and A is not equal to B. One cannot infer from this that A = D, and one cannot infer that B = D, but only that (A v B) = D.

88
Or perhaps he was only beginning to get through all the complication by the time he reached *De anima*. At any rate the discussion here seems better than the earlier one.

89
Aristotle admits that some parts do not help at 648a 16. He does not seem overly disturbed by this, presumably on the principle of "for the most part."

90
Preus (1975) discusses bilateral symmetry at p. 127. I am in essential agreement with him.

91
Thompson thinks otherwise linking three to Platonic mysticism, see: note to H. A. 513a 35.

92
Contrast to Aristotle's methods at 496b 5, 511b 15, and 515b 1.

93
Aristotle was aware of this problem but one wonders why he did not consider the nature of the heart valves as a means of defining the right auricle as distinct from the vena cavae. Even if collasped, the valve would still be there to demarcate it from a vein.

94
This question is discussed by T.H. Huxley, "On Certain Errors Respecting the Structure of the Heart Attributed to Aristotle" *Nature* 21 (1879): 1-5; Thompson, in his note on the passage, and by Arthur Platt, "Aristotle on the Heart" *Studies in the History and Method of Science* (Oxford, 1921), pp. 520-534.

95
De venarum arteriarumque dissectione, 9.

96
It could be that Aristotle had dissected a fetus' heart since in the fetus the ventricles are connected by a small opening which closes shortly before birth. This could account for the "common chamber" which in this case would be the left ventricle.

97
De fabrica corporis humani, liber 6, cap. 12.

98
There are textual difficulties with this passage which further hinder any definitive readings of the problem of the pulmonary artery based upon these lines.

99
This entire account is difficult and no single reading is entirely satisfactory.

100
There are also the appetitive and locomotive souls found only in certain animals (641a 23, 33ff.).

101
This property is interesting, especially when one remembers that it is cool blood that causes intelligence or extended sensation, P.A. 648a 2, cf. n.83.

102
Nutritive and rational souls are also lost in death. But plants feel the same temperature at death since they only lose the former. The only soul common to both animal and man is the sensitive. Therefore if both animal and man become cold at death, then it must be from losing the sensitive soul. The question is whether this argument builds the conclusion into the premises.

103
Food also causes heat, but the hottest part of the body is the heart (670a 20). See also Boylan (1982) for a discussion of heat and *pepsis*.

104
It is interesting to note how this definition has stayed with us in our legal codes causing much controversy lately concerning brain vrs. heart deaths, e.g., if the EEG shows no brain waves, is the subject still alive? See the ad hoc committee report of the Harvard Medical School that appeared in the *Journal of the American Medical Association* (1968) for an interesting discussion of the brain-death controversy.

105
The actual development of the chick takes twenty-one days. Seventeen hours after laying and incubation three germ layers of the embryo have appeared and the nervous system has begun to differentiate. By the second day the heart is beating and blood begins to circulate (Aristotle said the third day). The embryo's body becomes distinct from the yoke mass at three days when limb buds also begin to develop. By the tenth day definite organ systems have developed, the beak, eye and ear have formed and the growth of feathers has begun. By the nineteenth day the yoke sac has been completely absorbed and on the twenty-first a chick hatches.

106
Singer, pp. 24-28 talks about the historical significance of hierarchial formation of parts. Preus discusses epigenesis, p. 113, n. 6. Boylan (1983) discusses these in their historical context. Cf. Lones' reconstruction of the process, pp. 203-207.

CHAPTER 5

MODERN CRITICISM

There are two questions left to be discussed: 1) The relationship between the method and the practice, and 2) The adequacy of Aristotle's ME and TE in the light of modern discussion on teleology and functionalism.

An examination of the first of these will lead us into the second as well. Much has been written about the relation between method and practice in Aristotle's science usually, with the conclusion that Aristotle says one thing and does another.{1} This position (which I shall call the 'standard view') is correct in some of its conclusions, but mistaken in others. In evaluating this large question several smaller questions present themselves: a) Are the method sections of the Organon (particularly the Posterior Analytics) and the method sections of De partibus animalium 1 consistent with each other? b) Is the scientific method advocated in the Organon (particularly in APo.) carried out in the practice of the biological works, and c) Is the scientific method advocated in P.A. 1 carried out in the practice of the biological works?

These are difficult questions, but they expose the various aspects of the relationship between Aristotle's method and his practice in biology. Now the program of the Organon is varied and complex, but it is clear from the discussion in Chapter One that APo. 2.19 suggests a critical empiricism. One part of our answer, therefore, will be to examine how critical empiricism fares in P.A. 1 and in the biological investigations themselves. The standard view accepts the stated program of critical empiricism (in some form or other), but claims that in actual practice Aristotle did something else. Owen, in his article, argues that as a matter of fact Aristotle actually begins with opinions and creates dialectical arguments to prove his point

(Owen, p. 87). And though the bulk of Owen's analysis is directed at the *Physics*, he makes the same argument against Aristotle's biology. It is claimed that in the *Historia animalium* Aristotle appeals to common opinion:

> And of the biological "observations" many were bound to be hearsay, *legomena*, to be treated with caution. Such *phainomena* must be "properly established," ascertained to be "true data." In the same fashion the *endoxa* must pass the appropriate scrutiny, but in doing so they too become firm data. Nor, if Aristotle associated the *phainomena* with *empeiria*, as he does in the text from the *Analytics*, must it be supposed that his words are meant to apply only to *phainomena* in the first sense. *Endoxa* also rest upon experience, even if they misrepresent it. (Owen, p. 90).

Owen must be referring to those passages in the *Historia* in which Aristotle quotes other sources such as beekeepers as the origin of his data. Now there certainly would be a sense in which Aristotle would be relying on *endoxa* if he *merely* accepted the explanations of the beekeepers about some event. But Aristotle generally does not do this (with a few exceptions). Most often Aristotle accepts their first-hand reports of detailed descriptions. From these reports of their experience (accepted observation by experts) Aristotle can accept or modify the existing "reasons why"; or create new ones of his own. These accounts arise from the data in precisely the same way as Aristotle describes in P.A. 1 and APr. 1:

> I said earlier that we ought first to take the phenomena that are observed in each group and then go on to state their causes. (P.A. 640a 14).

> Consequently it is the business of experience to give the principles which belong to each subject. I mean, for example, astronomical science: for once the phenomena were adequately apprehended, the demonstrations of astronomy were discovered. Similarly with any other art or science. Consequently, if the attributes of the thing are apprehended, our business will then be to exhibit readily the demonstration. (APr. 46a 17-22).{2}

Indeed, with respect to critical empiricism, it seems that (a) presents no difficulty. But what about (b)? Is Aristotle actually engaged in uncritical rationalism (mere dialectic united to experience) or some other theory which falls short of the stated goals of the *Organon* and of P.A. 1? To point to

Aristotle's acceptance of the beekeepers and the like is not to show that Aristotle is solely relying on endoxa reports. On the contrary, the reason for relying on a consensus report is that one can have comparatively more confidence in "intersubjective" empirical reports than in novel ones. It is not common opinion that Aristotle is interested in, but common reports of descriptions of bee behavior. This is illustrated in Aristotle's discussion of bee reproduction (see 4.2). Aristotle rejects all common opinions about bee reproduction since they do not fit in with the reports given by the beekeepers about what bees do. Aristotle uses the reports to construct his own account. This account is consistent with actual bee behavior as descriptively reported to him by beekeepers. Aristotle is not here in violation of his espoused empiricism (first enumerated in the *Analytics*).

Duering makes a similar (though stronger) charge as he accuses Aristotle of being an "arm chair" biologist who speculated about living forms without any actual data to support his conclusions.{3} As was mentioned in the first chapter, one ought not denigrate the place of "rational" deliberation in scientific discovery, as the quotation from Einstein proclaims, but certainly there must be some kind of interface between these activities. For we want, in science, to make realist assumptions about our results. We will not accept *any* geometry as representing space (there are a number of very queer ones), but only those which purport to be about what is real (and, in principle, if not in practice, admit to some verification). It is this subtle balance which we have termed critical empiricism. And one need merely look at the examples given in the last chapter to convince himself that Aristotle indeed engaged in both activities (particularly in non-human cases).

Perhaps a stronger charge against (a) and (b) comes from those who point to the lack of syllogisms in Aristotle's scientific practice.{4} As Jonathan Barnes has remarked, "If the *Organon* were lost, we should have no reason to suppose that Aristotle had discovered and was mightily proud of the syllogism" (Barnes, p. 125).{5}

Solmsen, in an excellent article, examines this apparent incongruity and offers four alternatives.{6} First, one could reconstruct the arguments in the text into syllogisms. On this interpretation all Aristotle's actual investigations are syllogisms in disguise. But, on the face of it this alternative is unacceptable because it requires altering the text and because the relationship between syllogism and demonstration is not necessarily reflexive or symmetrical (i.e., that x is a syllogism and/or demonstration must be proven separately).

Second, one might contend that the rules of demonstration are softened to correspond to the practice.{7} This view would hold that the syllogisms of the the _Analytics_ are put together in a looser form in the practice to correspond to varrying problems. But since the _Analytics_ were meant to cover syllogisms it seems strange that no mention of a softened form or alternate syllogism is given.

The third possibility is that the range of applicability is solely to the rigorous sciences and that the only disciplines that qualify are: arithmetic, geometry, stereometry, mechanics, harmonics, astronomy and optics. Other, less rigorous, sciences deserve less rigorous modes of demonstration.{8} The difficulty with this argument is that it required biology and other "less rigorous" sciences to be non-necessary or contingent. (The evidence cited for this is the frequent expression _epi to polu_, for the most part.) But _epi to polu_ is not properly replaced by _endechetai_ (contingent). One could not, for example, use _epi to polu_ in the Sea Battle example. On the contrary, it is hinted that what is _epi to polu_ may be eventually known and capable of systematization (cf. APo. 75b 33-36, 98b 29-38).{9} In addition to this it is doubtful whether there is a connection in Aristotle between 'rigorous' and 'demonstrative.'{10}

The final interpretation sets out that Book One of APo. is really about presenting a finished science. This can be seen through the word, _apodeixis_ which means to show or reveal. The opening words of APo. 1 refer to teaching, "All teaching and learning comes from pre-existent knowledge." Thus APo. 1 is about how a body of scientific knowledge ought to be presented. For modes of discovery in the _Organon_ one must turn to APo. 2 and parts of the _Topics_.{11}

The analysis of (b) permits further exploration of (a). Now part of the problem rests in the apparent dissimilarity between the _Parts of Animals_ and the _Posterior Analytics_. Both the personae and the action seem different in APo. (which in its two books speaks about the transmission of an accepted science and of scientific discovery, respectively) and P.A. 1, which seems to concern itself with specific biological dilemmas. The reason for this difference in presentation by Aristotle lies in the respective functions of the two works. The APo. is a detailed treatment of scientific method in general while P.A. 1 is a specific inquiry into methodological problems dealing with animals. Because of its limited purpose, P.A. 1 only addresses questions cogent to the foundations of biological inquiry. Though P.A. 1 does not deal with the variety and range of topics that APo. does, P.A. 1 does follow from some of the principles laid down in the APo. These principles are mostly involved in 1) The kinds of questions that science asks, and 2) A mode of insuring against error by thoroughly grounding everything in a theory of definition which becomes the overarching tool for Aristotelian method. It is definition which allows critical inquiry to proceed in a systematic and rigorous manner through the mode of discovery.

These two modes will be briefly examined together. The basic concepts described in the _Organon_ all have their foundations in the _Metaphysics_.{12} The doctrine of the four causes which is introduced by Aristotle in _Metaphysics_ 1.3-7 find their expression in the _Organon_. Especially important are the formal and final causes. These find many expressions in the logical works. For example, the formal cause can refer to: _ti esti_, _to ti ēn einai_, _eidos_, _logos_, _horismos_, _kai ousia_ (7.4-6, 10; 1.3; 5.8). This wide range of meanings indicates that the close tie to "essence" (in all its forms) has with "being." Since it is the job of first philosophy to study being qua being (Gamma 1), and since the formal cause in its various forms is tied to being (enabling a large network of logical apparatus to be established), then the study of first philosophy is necessary in order to properly ground formal cause (and its corresponding logical network). This becomes clear in Z when Aristotle says at 1028a, "Now of all these senses which "being" has, the primary sense is clearly the "what," which denotes the substance." Book Zeta goes on to lay the foundation of the "what" including all the various senses of the word mentioned above. This kind of analysis is necessary for adequate clarity absolutely requisite for a logical exposition. It is this careful scrutiny which allows Aristotle to make the subtle, technical distinctions he does in the _Organon_. The intended scope of the _Analytics_ can best be appreciated by examining the scope of its foundations, viz., the discussion in the _Metaphysics_. It is here that ME and TE have their _archai_. The general metaphysical doctrines of _to ti ēn einai_ mold our sense of _eidos_ as form.{13} In the biological works it is this appreciation of the form or essence of a thing which makes conditional necessity and teleology possible. For these processes require an approbation of the form as it is: the being of x in _all_ of its various senses.

MODERN CRITICISM

The careful distinctions made in the *Metaphysics* allow Aristotle to be very exact when speaking about *definitions—ti esti* and *horismos*. Through these metaphysical questions *the physical schemata* of form and matter, actuality and potentiality etc. are also raised for analysis. As a result of these concepts, once clarified, it is possible to use them in a rigorous, consistent way. It is this rigor and precision which were absent before analysis and which are requisite in order to construct a workable logic and scientific method. It is the *Metaphysics* then which grounds the *Organon* and the *Organon* which *grounds all* of Aristotle's *biological investigations*.

It will be remembered that APo. 1 describes a mode of exposition. It is concerned with formalizing observations and explanations into a system of knowledge. P.A. 1 is not concerned with this formalization. Therefore, the carefully considered syllogisms of the former are absent from the latter. It is describing a developing science and how further inquiry ought to proceed. This does not, however, make it impossible to determine whether or not there are any significant inconsistencies with that part of the *Organon* which covers such developing sciences. Though *Aristotle does* not go into detail in P.A. 1 concerning the nature of a definition, he does use his previous analysis to make points in P.A. 1. In APo. 2.1 Aristotle asserts that a definition answers the question "*ti esti*." The definition states the essential attributes *of a subject* so that one understands the causes of a subject (i.e., all the causes which are applicable). Aristotle uses this as he criticizes other *phusiologoi* who mistake the accidental for the essential.

> So Empedocles was wrong when he said that many of the characteristics which animals have are due to some accident in the process of their formation, as when he accounts for the vertebrae of the backbone by saying "the fetus gets twisted and so the backbone is broken into pieces." (640a 20-22).

The source of Empedocles' error is that he was ignorant of certain basic biological facts, viz., that the *eidos* of a *zōon* is predetermined through its *logos*.{14}

This statement needs some unpacking. The *eidos* of a biological form is not just its shape (*morphē*), but *its shape* in conjunction with its function (P.A. 640b 34). Now both the shape and those functions can be empirically determined by the biologist. A segmented spine which has uniformly shaped vertebrae can be ascertained through dissection. Why is there such regularity? This is because the spine has a certain set of tasks that have to be performed (653b 35, 641a, cf. De incessu 702b 20). A non-segmented spine could not perform these functions. If these tasks are necessary to survival of that species, then the biological form must have a segmented spine. But what about an irregularly shaped spine? Here again if the spine were broken irregularly by twisting and turning of the fetus, then there is no assurance that the individual could perform the tasks necessary for being a member of that form (i.e., being able to survive as a such and such). It is because certain tasks need to be performed to insure survival in some form's ecological niche that particular parts are structured in such and such a way. The eidos of a form allows the biologist to understand the part in question more completely. Both the part and its function are revealed to the biologist. The spine as a matter of fact does allow certain tasks to be performed: bending, running, twisting, fighting etc. Thus there are two levels of justification for a spine being segmented: 1) By shape—one can notice through dissection that animals do have regularly shaped vertebrae. This regularity in an end product suggests a regularity in the process which produced such a product. 2) By the function of spines. In order for animals to fulfill their functions (i.e., to live and reproduce) they must perform certain physical activities. These physical activities demand a spine which is regularly segmented, thus the spine must be regularly segmented. And if this segmentation is so important to the species, then it would seem odd if this formation were relegated to chance. And besides, if it were, then one should expect to see many individuals of that species with irregularly shaped spines which cannot perform the requisite functions. We do not see these in the number that chance would statistically require, so we can infer that the spines are not formed by chance. At this point it is interesting to examine a few proponents of Empedocles' viewpoint in the contemporary arena. These are the detractors of teleology and functionalism. Most of these criticisms can be lumped into three groups: 1) Teleological explanation implies a conscious agent at work;{15} 2) Teleological explanation excludes necessitating causes;{16} and 3) Teleological explanation is not genuinely explanatory.{17} Now the first two of these objections have already been dealt with. In Chapter Three it was shown that Aristotle need not have personified Nature, but could have adopted a "systems" approach. Indeed, it seems unlikely that Aristotle did view teleology as implying a conscious agent. Further, through the systems approach outlined, it is clear that Aristotle is not committed to a conscious agent in any real (i.e., non-metaphorical) way. The second objection was solved in the same chapter through an examination of Aristotle's ME. ME is a conditional necessity which has teleology built into it so that the problem of incompatibility vanishes. If Aristotle had maintained simple necessity and teleology simultaneously, problems might have occurred, but in phusis this is not the case.{18}

It is then the third objection which will occupy most of our attention. This claim takes on several guises, but all of them revolve around the insufficiency of teleology to explain some event. Now it is clear that what I have been promoting might be termed "functional" or "systems-property" accounts.{19} As we have already shown teleological explanations, in which the conscious purposive agency of the individual or system (except in some metaphorical sense){20} is intended, are not pertinent to our analysis. These systems are not conscious and so when we speak of functions it is in terms of the work (ergon) to be performed.

One of the more eloquent spokesmen for the position that teleological explanations (functional explanations) are not truly explanatory is Carl Hempel. In his deductive-nomological model (d-n), he explains that some event (E m+1) is explained by the conjunction of (E 1 . . . m) and the appropriate covering laws (L 1 . . . x).{21} These "covering laws" subsume the explanation.{22} For Hempel the explanans is the set of preceding events and the covering laws which explain the event in question.{23} This indicates the positivist tie with prediction, which is generally yoked with explanation to create scientific theory (Prediction + Explanation = Scientific Theory).{24} Such an account of scientific theory is controversial, {25} and it is clear that since (at least if and until it is a finished science) biology (for Aristotle) can only generate epi to polu conclusions; it is not primarly interested in prediction. Indeed, Aristotle rarely, if ever, indulges in this sort of speculation. For my own part, I believe that prediction is a useful tool for biology and other sciences, but that its principal formal role is in establishing a body of scientific data, providing its empirical verification and aiding in its implementation.

Hempel objects to statements such as "the function of the heart is to pump blood to the body" because of difficulties in the form of such a statement. The preceding functional ascription translates into "F(i) in organism S during time t in environment C is to do n," which further fits the statement, "i in S during t in C has effects e that satisfy n and are necessary to S." To make this an explanation Hempel puts three requirements:

1. During t in C, S is properly working.

2. If S is working properly, then n will be satisfied.

3. If i is in s, then effect e of C's presence satisfies n.

But if this account is correct the assumed premises do not entail the conclusion, therefore; there is no explanation. One could substitute a 3' saying, "condition n is satisfied only if item i is in S." But this, Hempel believes, is in general unwarranted, therefore; Hempel believes that functional accounts may have heuristic value, but they have no value as explanations.

These problems have been raised in differing guises by others but their resolution requires: 1) A clearer notion of what is scientific explanation, and 2) The proper role of 'necessary' and 'sufficient' conditions within teleological explanations. To answer the first point Aristotle leads us in the right direction. He states the general form of scientific inquiry in APo. 2.1 as getting the fact (hoti) and the reasoned fact (dioti). The dioti is embodied in TE and ME. Put in more familiar terms: Explanation is an account which helps us understand some event through the relevant operating systems.{26} This includes two modes: a) the 'how' and 'why' of the systems' operations, and b) the necessary operations of the systems' parts. One should notice that this model of explanation is quite different from Hempel's. An examination of a few examples will illustrate this. One objection often given to the d-n account is that it has difficulty explaining statements of the form, "why does x have phi?". Take the question, "Why is this bird black?". The d-n account would construct an argument beginning with the law "All ravens are black" and then subsume the particular, "This x is a raven" to get the conclusion, "This x is black."{27} This, of course, is subsumption and not explanation. According to the above definition explanation requires understanding of some event through the relevant operating systems.{28} In order for the d-n practitioner to do this he would have to isolate that which is responsible for the color being what it is. This requires an account closely tied to the object of explanation and not a mere concomitance of ravenhood and blackness. By focusing on raven as a system, one would be forced to observe its inputs and outputs (activity) and, from such "black box" analysis, construct a reasonable account of how and why blackness contributes to the system's operation.

Now the d-n account's charge against functionalism that 3' is needed to make the deduction work is countered in two ways. First, 3' is basically the requirement that given P => Q that if -P then -Q. But this, of course, is not true of material implication because the model F => T is true. Counterfactuals, of course, have proven to be equally troublesome for the d-n advocates themselves.{29} It is an odd attack to say that some other account is to be rejected because of psi when one's own theory also possesses psi. But secondly, the model of explantion I have just offered as a free reconstruction of Aristotle does not need to rely upon a deduction model containing material implication (and in the developing science no deductive model needs to be applied as has been mentioned above).{30} Such logical paradoxes need not worry the developing science, but are only encountered by the "completed" science ready for transmission. However, even here counterfactuals might be explained through an emphasis upon necessity found in ME along with the reversed order of antecedent and consequent. (A more complete examination of this question mus remain a task for another essay.)

In order to show how ME and TE function as explanations the proper roles of 'necessary' and 'sufficient' conditions must be illumined. TE explains the why of a system's structure. It aids in our understanding of the reasons which justify the ordering of the system and its operation. Now this mode is often found to be confusingly vague. One case in point is the "use/design" distinction. It would seem that TE requires some sort of "design" for without this constraint one cannot distinguish true purpose from fortunate accident. One might put a newspaper under a door to prevent drafts, but one could not say the newspaper's function was to prevent drafts.{31} Another example is the case of a small oil drip which keeps a timing mark clean on a car. Just because some happening, x, is useful, does not mean x has a function. To be purposive requires more. When I go to the agora to buy some fish and meet a man who owes me money, that is a happy accident. X is fortuitous, just as in the last example. In the case of the car it is a "sport" or mutation. If this accident proves useful I might take out a patent on such small holes and sell it to Detroit (or Tokyo?). If the sport is useful it will be repeated because of its positive fitness and it will soon predominate. As it begins this process, it acquires a function. The "function-ness" is the result of other established functions which order the coming-to-be and passing- away of lower level functions. This suggests a hierarchy of functions which corresponds to the systems they represent. These systems form a nested order. Each system is discrete as are the parts which compose the system. Those parts may also be systems moving down in specificity indefinitely (though there is a point in which one passes from distinctly biological systems into chemical and physical systems respectively). Since the parts are robust they can be examined independently in order to ascertain their natures and capacities.{32} And because they form systems, each part interacts with other parts—both on its own level and on other levels to help bring about a specified output. This is not an affirmation of the unity of science{33} but merely a description of the interactions of robust systems located in an ever increasing hierarchy of other systems.{34} There are two types of systemic interactions: those on a single level and those between levels (a level is a strata within the nested hierarchy). The former are horizontal interactions and the latter are vertical (+ equals a vertical affect upon an upper level; while - indicates an effect on a lower level; 0,+ and 0,- equals the two being affected respectively).

When one wants to determine the function of some system it is necessary to examine the horizontal and the vertical 0,+ and 0,- interactions. This involves asking of the output x, "What is x good for?" or "What does x do?".{35} This question focuses upon a single function for a given event. To say that the function of x is y is not to say that under a different description (a different event) there might not be another function to x.{36} This is because TE is neither sufficient nor necessary for any given outcome. Remember that for TE to be explanatory it needs to help us understand the relevant operating systems. Certainly knowing what x 'does' or 'is good for' tells us a great deal about the operation of those systems. It provides the <u>why</u>.

The <u>how</u> and the necessary operation of the parts are all that <u>remains</u>; this constitutes the ME. It should be remembered also that ME is <u>conditional</u> necessity. This creates a statement of the following form, "In order to achieve a specified effect, e, <u>ceteris paribus</u>, i must be in system S during time t in environment C." The resulting conditional also holds, "If e has occurred <u>ceteris paribus</u>, then i is in S during t in C." I also define 'sigma E' as "all the relevant effects of S" ('relevant' here is defined with reference to TE) and ce as a proper subset of sigma E. From this we can say that in a single event sigma E or ce is the end or purpose of S. What is created is a conditional relationship with the effect as an antecedent over a specified class of interactions under a single description.

A: $(\exists y) \{(\text{sigma } Ey + cEy) \Rightarrow (\exists x)(\exists i)\{[(Sx \,\&\, TCy \,\&\, y=x) \,\&\, Fi] \,\&\, -Px\}\}$

[TCy means y has a specified time and environment; Px means x is subject to permuting factors; Sigma Ey means y has all relevant effects of S; cEy means y is a proper subset of sigma Ey; & is conjunction and + refers to exclusive disjunction.]

(A) differs from Hempel's formulation of how a function is to be characterized in several important respects. First, what (A) describes is an ME relationship and not a TE. Second, it focuses upon the effects as antecedent.{37} Third, (A) only refers to the effects E and does not refer to purposes n. This is because n is given by TE which conditions the outputs that shall result from S. Fourth, the <u>ceteris paribus</u> clause excludes possible outside permutations, for if Px becomes significant (one can assume that in most cases it is not), then the operation of S would need to be altered to S' which could account for said variation. This would result in a corresponding statement A'. However, in actual cases S and S' would most likely represent different systems. S' would be a new system which varies in a most significant varible from S. As S' systems became to be produced, gradually all the S systems would be phased out. This is a rather rough and ready example of natural selection and focuses upon an important distinction between ME and TE: plasticity. TE is plastic and persistent in that its purposes (especially at sufficiently high levels) can be achieved in numerous ways and if thwarted in one direction it can seek satisfaction in another. This can occur because the antecedent conditions of TE are neither necessary nor sufficient. In ME, on the other hand, there is a fixed directional component. Certain conditions are necessary and are sufficient, <u>ceteris paribus</u>, to produce e. This can be illustrated as follows.

Figure 7d illustrates the plastic, persistent relationship between a purpose, P, and an output, B, which satisfies that purpose. However, in 7a, 7b, and 7c the output is assumed to be the purpose of x. There is no variation in purpose. In 7a X is sufficient and necessary for B. In 7b the relationship is more complicated; X is necessary but not sufficient for C (i.e., B is also needed). But if both B and C are included in sigma E, then X is still necessary and sufficient for the output. In 7c the stated relationship between X and B, C, and D requires the addition of another factor, p, in order to decide between these three alternatives--if there is a unique e (in which case "cause" here is neither necessary nor sufficient), but with the addition of sigma E as the goal this difficulty is removed.

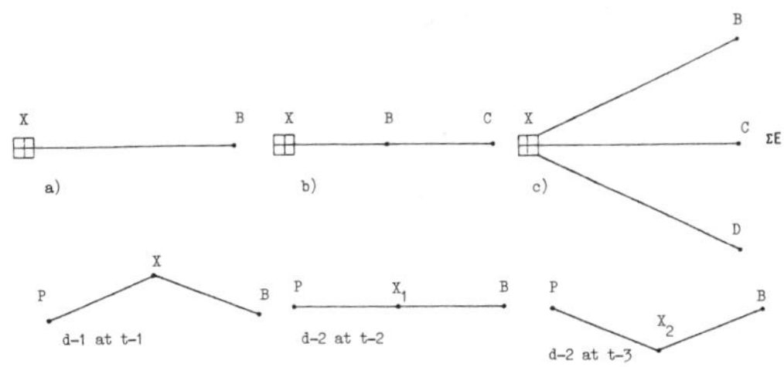

FIGURE SEVEN: Plastic and Directional Systems

Now it is important to read these ME figures from right to left, i.e., starting with the effects as ME demands. In this way one can note how ME is necessarily directional as contrasted to TE's plasticity. When Px becomes significant, say in an organism's ability to maintain a requisite body temperature in a changing environment, then the output of that system ceases to coincide with an upper level TE (such as those supplied by natural selection and biological heat requirements for that species), and those systems will die out. If there is sufficient diversity in the sample space in which the variant homeostatic systems for heat control can create an output more in accord with that TE, then they will thrive and become dominant. Thus it is that TE acts to condition and guide ME for by its persistent and plastic seeking of goals it sets the parameters by which groups of systems thrive or perish.

A second objection against teleological explanation is that it is really reducible to a non-teleological, mechanical one.{38} Nagel, for example, contends that teleological arguments may all be restructured into non-teleological ones, "The content of the teleological statement is fully conveyed by another statement that is not explicitly teleological and that simply asserts a necessary (or possibly a necessary and sufficient) condition for the occurence of a certain trait or activity of the organism" (Nagel, p. 405). This translation is given in terms of certain spatial and temporal components, "The [biologists'] contrast between structure and function is evidently a contrast between the <u>spatial</u> organization of anatomically distinguishable parts of an organ and the temporal (or spatio temporal) organization of changes in those parts" (Nagel, 1961, p. 426). What this amounts to is taking some teleological formulation and conveying it in this "purely descriptive language." An example Nagel gives is homeostasis. On the face of it homeostasis seems to be a good example for Nagel. The teleologist claims (in the case of homeothermism, one homeostatic system) that certain heat regulating mechanisms exist to maintain a proper temperature for the body. This can be rephrased by fixing the spatial components involved in a particular act of temperature regulation and describing the process of regulation step by step. From this description one can formulate a quasi-law-like statement such as Px => Qx where Px represents the conditions which bring about Qx (just as in the d-n account described above). It is clear that such formulations are intended to render particular biological events as instances of a law-like statement on the model of physics. One can point to several problems with this approach as an attack on teleology: 1) Biology cannot be reduced to Physics or laws on a physical model (as Nagel's formulation suggests), 2) The simple reduction proposed by Nagel is not amenable to biological systems whose interactions are on many levels and 3) Teleology as I have presented it is both a) the only way one can understand a biological event and b) the way that the event actually takes place.

First, Biology cannot be reduced to Physics. The unity of science hypothesis, so attractive to the logical empiricists, is problematic because biological questions are multi-leveled. As stated by Putnam and Oppenheim, the unity of science requires discrete boundaries between levels and linear structuring of the levels.{39} This first point has been shown by Paul A. Weiss to be an incorrect account of biological levels.{40} This point is reinforced by William Wimsatt and David Hull, who demonstrate that the simple reductionism, required by the positivist model, is not possible in biological systems.{41} This is because of: 1) Feedback possibilities, and 2) Upper level determination of lower level events, 0,+; –. Feedback is a phenomenon which belies the normal linear structure by allowing some level to inform another in a direction contrary to the "normal" direction of the event. For example homeothermism is just one of the many systems of homeostasis in which the direction of causation is not merely from lower to higher. When body temperature rises, a nerve center in the brain sets in motion a certain cooling mechanism or conversely sends out stimuli that increase muscle tension and thereby the amount of heat produced when the temperature falls. This balancing mechanism is quite complex, but it is clear that both upper and lower levels affect each other so that causal arrows would go in both directions (+; -; 0,+; 0,-). Biological feedback of this type confounds the simple notion of upward causation inherent in the linear structure. An example of the upper level determining the lower level can be found in evolutionary theory. According to most versions of evolutionary theory the phenotype's environment affects the genome. Thus a higher level affects the lower level.

All of the above examples are meant to demonstrate the integrity of the biosystem and each of its sub-systems. These afford a more accurate account of biological activity than reductionist theories do. But perhaps anti-teleologists might be willing to admit the integrity of systems analysis, yet still contend that if these systems are merely described chronologically as performing certain actions teleology is eliminated. This is true. The only way to eliminate teleology is to start at t1 and describe some event which has an end at tn. This would eliminate teleology and this is the thrust of Nagel's intention—yet what good is such an account? First, under such an account one would be excluded from knowing what was happening in the case of a developing spine. One could allow knowledge of the atomic tables perhaps (though this also is controversial), but what are these elements doing? What is the significance of each bonding and the spatio-temporal events? None of this can be known since on epistemological grounds the observer does not know what is being formed. Indeed, all the events distinctive to "spine formation" could not be understood since being novel they are purely a function of "spine organized as such and such." The significance of any of these steps can never be understood until one knows what is being formed. It is only after this naive observer has watched the entire process that the elevance of the events preceding can be known. It is the end which makes the process comprehensible.{42} This does not merely mean that TE is heuristically useful, but that it is necessary for our understanding of how and why an event occurs via the relevant systems, and this, as we have said above, is exactly the nature of scientific explanation.

Secondly, this is the way biological events actually take place. Every determinate outcome must have a regulative principle or mechanism. In a system such regulatory mechanisms are not capricious, but must work together to produce some upper level effect. If each of these subsystems were not regulated by some upper level principle (0,+ interaction), the whole would be completely divisible into the sum of its parts. But this has been empirically shown not to be the case. A ciliate protozoan shows that the rows of cilia (which are systems) are ordered in a pinwheel fashion around a central core. The sum of the variation in the cilia individually is greater than the variation in the rows themselves.{43} The same holds true for the arrangement of mitochondria in a sperm's tail. These two examples show that the variance, v(a) or a part A and of other parts v(b) to v(n) is less than the variance of the total system, V(s). $V(s) < \Sigma\ [v(a) + v(b) + \ldots v(n)]$.

Why is this so? Obviously there is some emergent property at the higher level which is expressed via that level. To ignore the higher level in scientific explanation (which is what anti-teleologists such as Nagel would do) is to ignore an empirically verifiable mode of how organisms actually develop. Biological systems, being regulative in nature, must develop teleologically. Thus, it has been shown that organisms: 1. can only be understood via teleological explanation and that 2. bio-systems do, in fact, develop that way.

Aristotle does not take his argument in quite these directions, though none of these arguments are inconsistent with the doctrines of TE and ME which have been outlined. Perhaps now it is appropriate to return to Aristotle's attack on Empedocles' account of the spine's development.

It is interesting to notice in detail the presuppositions necessary for Aristotle to generate this argument. The definitions must be established carefully. Aristotle says of Empedocles in P.A. 640a 23: "He was unaware first that the seed which generates the animal must initially have the appropriate specific character and secondly that the producing agent was pre-existent both logically and chronologically." These two points rest upon the implied axiom that one is concerned with in generation: 1) what is to be (cf. 640a 3) instead of what actually is, and 2) the fulfillment of this future via ME and TE (see 3.2 all of 3.3 and Chapter Four). These devices of analysis are axioms, viz., incapable of justification such that they must be accepted before any further discourse can proceed (definition of axiom, APo. 1.2). Among the axioms there are at least two varieties: a) axioms common to all science such as the laws of identity and of contradiction, 2) the axioms peculiar or proper to a particular science or sciences as in the geometrical axiom, "A straight line is the shortest distance between any two points" (Metaph. 4.3).{44} Therefore, the important metaphysical doctrines already discussed, such as TE and ME; form and matter; and actuality and potentiality etc. are all axioms which seem to be between these two varieties

(Aristotle does not discuss the question regarding TE and ME as axioms--perhaps this is because biology is not yet a finished science and ready for transmission). They are not as broad certainly as the axiom of non-contradiction nor are TE and ME applicable to all sciences. However, they are propositions which cannot be debated, since they are requisite for further discourse and they deal with more than one science. Perhaps one should classify TE and ME between these two large groups. Aristotle never demonstrates positively that the four causes are necessary and sufficient, but rather gives indirect proofs for them being separately inadequate: e.g., Material alone, P.A. 640a 20, 641a 10; Phys. 193b 5; Metaph. 1029a 2, 1026a 5; Efficient alone, P.A. 640a 20, 641a 10; Phys. 2.4-7; Metaph. 1065a 30; Final and Formal and their relation to ME, P.A. 641a 25, 641b 20; Phys. 2.8; Metaph. 994b 10. None of the causes singly is adequate (just as neither ME nor TE alone is sufficient). This cannot be shown positively (i.e., through demonstration) but can be made plausible via dialectic (cf. Top. 1.2).

Now to arrive at the subject, in this case "human segmented spines," one must prove that certain attributes attach themselves to this subject essentially. From TE one knows that if a spine is to be segmented (the necessity of which has already been shown), then it must display some purpose. Aristotle does not go into the particular functions of the spine here, such as its usefulness in bending and stooping (among other things), but he does mention that a spine is necessary to carry out the function of Man 640a 35. As a matter of fact, a man _does_ have a spine and it is part of his _eidos_.

> It would be better to say that because Man is the way he is (with such and such body parts) that because of this he has this form (containing those parts). For if he is without these parts it is either impossible for him to exist without them or at least that it is better that they should be there. (640a 35).

There are certain parts perhaps that a man could do without, e.g., an arm, a leg or an ear, but a man is not whole without them, i.e., does not have the complete _eidos_ of Man (or more generally, of Blooded Animal). Other parts are necessary in the full sense, e.g., a heart, a brain--and though it is not mentioned explicitly as such by Aristotle--a spine. The spine has a particular shape and a singular set of functions which can be determined through careful observation and dissection. This is necessarily seen to be, therefore, a part of the _eidos_ of

Man. But could not "spine" be shaped by accident and still be a part of Man's eidos? Could not spine simpliciter be a part of Man's formula and not "spine shaped in such and such configuration?"{45} No, says Aristotle, because it is the nature of TE that it convey some purpose. That purpose requires a plan. Just as a statue maker must devise a careful plan to execute the exact duplication of his artistic preconception upon some slab of marble, so also must the plan of a spine have its own precise potential execution. A statue cannot be created through chance and neither can a spine for if this were the case one would expect a sizable variation in spines. They would not be what they were necessarily. That is, there would be no formal or final cause. We can empirically verify that men have segmented spines. Now if a man has 'segmented spine' as a part of his logos, then the spine must be produced in the same way each time. The process of formation and the formation itself, its order and outcome, must be carefully regulated.

> Because man is such and such, therefore the process of his formation must of necessity be such and such and take place in such a manner; which is why first this part is formed then that. (P.A. 640b 1-2).

The order must depend upon the purpose. In other words, Aristotle is employing TE.

ME is also involved: "that the producing agent was pre-existent; it was chronologically earlier as well as logically, in other words it was begotten by men" (640a 23-5). The argument states simply that if one is to form a spine in such and such a way, a determinate process is needed. Chance does not afford such determinism because it may be different every time. There needs to be a specific chain of material and efficient causes to produce the segmented spine. Therefore, chance cannot be the cause (in any sense) of a segmented spine.

The attribute in question is the segmented spine's mode of necessary determinate formation. It is shown to inher essentially within the subject "human spine," which is a necessary part of the eidos of the subject genus "Man," via the modes of TE and ME. Aristotle needs this machinery for definition in order to produce his conclusions. When considering some part of a species, one must first determine the connection between the part and the form of the species. Once this is determined the species can be examined in terms of the genus so that finally a relationship between the part and the genus can be ascertained. This is to allow knowledge of the part at all relevant logical levels. This knowledge is necessary in order to determine the status of the part in question and can thereby be put to general use. For example one might ask what the consequences of hectocatylization are (see 4.3). This question would be phrased relative to the large genus of "molluscs" and to the genus cuttlefish (species: Argonauta or octopod cuttlefish, H.A. 524a 4-5, 541b 9-15, cf. 544a 12; G.A. 720b 33). So it is in other biological examples. The order of Aristotle's inquiry follows the logical order established through his theory of definition set out in the _Organon_ and based upon principles found in the _Metaphysics_. This connection between the part and the subject genus in the example of the human spine (i.e., the spine and its relation to Man) has already been examined at 640a 35. It seems that Aristotle's point here is a logical one: the essential definition of the subject genus is the sum of all variations in intension within the subject genus' extension. This appears, at first glance to be phenetics (cf. numerical taxonomy in Chapter Two), but really it is closer to cladistics in the special sense that was outlined earlier. The human spine is a generalization created through imperfect induction, i.e., an examination of a number of individual spines. From this examination one knows the standard deviations in spines within the sample space. Within these parameteres a description of human spines can be constructed. Now if the particular spines examined were determinate, then so also will be the necessary part in its relation to the subject genus. Further, if a group of end products have a determinate _eidos_, they must have been produced in the same way, ergo the _process_ must have been determinate and not accidental as Empedocles asserts. This argument is about biology and yet the methodology is derived in part from Aristotle's general work in the _Organon_ both regarding the general questions to be asked and a method of definition which will allow one to proceed in the investigation.

The key to Aristotle's rejection of the Empedoclean "chance" axiom rests in the paucity of its account. It could not supply more than the _hoti_. This was also the problem with Anaxagoras (640b 6). They only give an efficient or a material cause which ignore the fact that all organic parts exist as organized within a system that has definite functions. In the example of the statue's cavity (641a 10), a material explantion of the cavity would be, "that the cavity is wood." This does not tell us much about the cavity. If one asked the artist, "Why is there a cavity in the statue's cheek?" an answer that the cavity is made of wood is very unsatisfactory. Likewise, the efficient cause alone would force one to respond, "It is there because of a certain kind of motion from my chisel." Thus material and efficient causes alone do not convey enough information. It would be much better to know that the cavity was created to depict a gaunt face in order to demonstrate the pain of hunger (for example). This addition of the formal and final causes makes physical and biological explanations more complete. By giving only two causes at best these competing theories fail to account for the relationship between the structure of form of a thing and its function. Indeed, the _dioti_ requires not only a more complete enumeration of causes, but of proximate causes. When one only offers an explanation of the fact, the asserted cause can only be remote (for a discussion of these terms see 1.5). It is interesting to note Aristotle's use here of _hoti_ and _dioti_. This distinction which is introduced in detail in APo. 1.13 is taken up in P.A. 1, 641a 15:

> It must be evident that the statements of the _phusiologers_ are not satisfactory. We have to state how the animal is characterized, i.e., what is the essence and character of the animal itself, as well as describing each of its parts; just as with the bed we have to state its form.

The _dioti_ or "why" is missing from the other natural philosophers. Without this knowledge is impossible.

Again, it is not the case that Aristotle is interested in presenting all the technical details of this distinction _per se_ as he did in APo. 1.13. Rather, Aristotle is using a distinction he has already made elsewhere for his particular purposes here. It is not necessary in P.A. 1 to explicate the _per se_ distinction at length, but rather he uses the distinction as already given. It is this very usage Aristotle employs when he says in P.A. 1 641a 10:

> The artist's [account] will be better for it is not enough to say "the fact that" (_hoti_) after the tool struck, this became hollow and that flat, but he will give the "reasoned fact" (_dioti_) or "why" he made the blow so.

The fact of the cavity is not enough. The reasoned fact is also requisite.

In the same manner Aristotle follows his work in the _Organon_ when he criticizes the methods of classification by other writers (particularly the practitioners of dichotomy; see discussion in 2.4). Many of Aristotle's points made in P.A. 1

on classification follow his discussion in the Topics. In the Topics Aristotle is dealing with the general problem of definition by division whereas P.A. 1 speaks to division only within the biological realm. This can be noted in the discussion on privations (cf. also the discussion of division in APr. 1.31; APo. 2.5,13).

> If the opposite of the species is a privation, we can demolish an argument in two ways, firstly by seeing whether the opposite is found in the genus assigned; for either the privation is found anywhere at all in the same genus or not in the ultimate species . . . secondly, if a privation is opposed both to the genus and to the species, but the opposite of the species is not found in the opposite of the genus. (Topics 124a 35-37).

> Again, this method of twofold division makes it necessary to introduce privative terms, and those who adopt it actually do this. But a privation, as privation, can admit no differentiation; there cannot be species of what is not there at all . . . . Now it is difficult enough to arrange the various animals under such lines of differentiation as these, which after all do contain species, in such a way that every animal is included in them, but not the same animal in more than one of them. (P.A. 642b 20-23, 30-32).

The first passage is more concerned with proving a logical point while the second accepted the logical point as already made and proceeded to make a biological application. In this way Aristotle is following the method laid down in earlier works and applying this to biology (cf. P.A. 644a 1-12 to Top. 144b 3ff.).

Another example of an argument made in P.A. 1 which follows from Aristotle's earlier work is found at 643a 27:

> Again the division ought to be made according to points that belong to the essence of a thing and not according to its properties (inseparable attributes). For instance, in making divisions of geometrical figures, it would not be right to divide them into those whose angles are together equal to two right angles and those whose angles are together greater than two right angles; because it is only a property of the triangle that its angles are together equal to two right angles. (P.A. 643a 27).

> 'Accident' has also another meaning, i.e., all that attaches to each thing in virtue of itself, but is not in its essence, as having its angles equal to two right angles attached to the triangle. (Metaph. 1025a 30).

> Of the accidents that are not essential according to our definition of essential there is no demonstrative knowledge; for since an accident, in the sense in which I am now speaking of it, may also not inhere, it is impossible to prove its inherence as a necessary conclusion . . . it follows that we cannot in demonstrating pass from one genus to another. We cannot, for instance, prove geometrical truths by arithmetic. For there are three elements in demonstration: 1) what is proved, the conclusion--an attribute inhering essentially in a genus; 2) the axioms, i.e., axioms which are premises of demonstration; 3) the subject genus whose attributes, i.e., essential properties are revealed by demonstration. (APo. 75a 18-22, 39-42, cf. 76b 11-16).

This discussion concerns the relationship between the three terms of a demonstration. When a connection is essential, then the demonstration is possible (assuming the premises are true). Further, when deciding upon the subject genus one must examine its essential attributes. In order to properly predicate an essential attribute of a subject genus, there must be a middle term which may be predicated of the subject genus and of which the predicate of the conclusion may be predicated in the major premise. It is because of this that the subject genus must have a univocal intension and extension. One cannot mix axioms between sciences which are peculiar to that science. This is because each subject genus refers to a unique set of propositions which describe it. This set of propositions marks off the domain of that genus. To mix genera would be to make a serious error for it would attribute essential properties of one subject genus to another, which by definition, is impossible.{46}

This emphasis upon discrete, "natural" groups is carried over in other ways to biology. In fact Aristotle's entire systematics is based upon this notion of establishing "natural groups" which are marked off by essential characteristics (determined by TE and ME; see my discussion in 2.5). Thus 'bird' cannot be grouped according to 'flier' since there are several different animals which fly (such as insects). Also this would not include birds which are non-fliers: those which live on land or on water. The attempts by other natural philosophers to make such divisions do not succeed because they fail to mark off the natural groups. They are not identifying the essential attributes, but instead mistaking accidents for essential attributes.{47} Aristotle's whole systematics and his criticisms of other attempts at classification are based upon his work in the _Organon_, particularly in the _Topics_ and _Posterior Analytics_. The distinction he draws between various large genera, the small genera and the species employ the notions of class inclusion which Aristotle developed earlier in the _Metaphysics_ and then in the _Organon_ as modes of definition. As opposed to division there may not be a single differentia, but rather several which mark off a group. The group itself must be understood as having some essential attributes. Thi search is really one of comparing the _eidos_ of one species to that of another. It is an empirical process of employing perfect and imperfect induction. Here as elsewhere Aristotle has taken a general logical point (consistent with his thoughts on metaphysics or being) and uses it for a specific scientific purpose. Class inclusion and his ideas about essential definition make it possible to construct a systematics and a taxonomy wherein all _zōon_ may be classified.

Thus it has been shown that some of the most general tools of analysis such as: the causes, the form/matter distinction, actuality and potentiality, and the four scientific questions of APo. 2.1 (_hoti_, _dioti_, _ei esti_, and _ti esti_--the first two by themselves; the third in conditional necessity and the fourth in a general theory of definition) are all included in Aristotle's philosophy of biological science (a) as well as being carried out in actual biological investigations (b) and (c) (cf. the questions posed at the beginning of this chapter). It is also true that there is much in the _Organon_ and the _Metaphysics_ which is not actualy employed, but by and large this is because these regulations refer to the transmission of a finished science and not to the discovery phase of critical empirical investigation. One can find in a mature work like the _De generatione animalium_ the same guiding devices for discovery which were undoubtedly laid down years earlier.

Aristotle was not always successful in carrying out his critical empiricism in (c). It will be remembered from the last chapter that there were frequent lapses from his espoused method when he actually engaged in biological investigations. These were perhaps most apparent in his discussions on birds, sea urchin ova, and on human anatomy (particularly on the brain). In these cases there was an uncritical reliance upon the _endoxa_. The result was wildly wrong conclusions based upon a faulty utilization of his own methodology. As indicated through our detailed examination of the practice Aristotle's results here are mixed. But even though Aristotle does fail to follow his method at times there are other instances in which Aristotle was very successful. His astute observation was responsible for his remarkable description of hecotcatylization and of insect anatomy. And it was as a direct result of his method that he was able to account for bee reproduction, sleeping fish and the cleidoic and non-cleidoic eggs (especially the so-called expanding eggs in fish). These accounts are not small achievements. They combined sharp observational skills with a careful methodology which produced good results.

There were also times when the results were wrong, but the method which yielded them could not be faulted. For example, Aristotle's argument about the body temperatures of insects rests on the assumption that insects have no blood. This assumption was made because insect blood is not red. Therefore the fluid which Aristotle could observe within insects could not be blood. This conclusion, though wrong, cannot be blamed upon Aristotle because lacking more precise equipment for analysis of blood (such as a microscope and/or chemical testing equipment) his result was the natural one to be made. Or again when Aristotle cracked open eggs to determine which organ was first to develop, he concluded that it was the heart since it was the first one he could see, and he built his theory up accordingly (resulting in several quite inaccurate conclusions). However the observation and the method Aristotle employed cannot be faulted. That he lacked the staining techniques needed to detect the spinal cord and developing central nervous system was not his fault, and one has to admire his technique and observations even if they brought him to inaccurate conclusions at times.

What we are left with is a biologist who, to be sure, made mistakes--both in following his method at times and in drawing correct conclusions at others--but also one who seemed, on balance, to be concerned with following his biological method. We have a biological method generally grounded in a rigorous philosophy of science that is critically empirical. And even though there may be some lapses here and there these should not blind us to his achievements which are considerable both in method and practice.

It is perhaps because Aristotle spent so long working out general methodological considerations (philosophy of science) and specific applications of method (as in his philosophy of biology), that Aristotle was as successful as he was in his own biological investigations. It was from carefully worked out analysis of the *archai* of the disciplines involved that his method and his strong commitment to critical empiricism evolved. This methodology and its application are thus the end result of a long investigation which was no accident, and Western Science, particularly biology, owes an inestimable debt to this innovator of scientific exposition and discovery.

Notes to Chapter Five

1
This view is taken by Jonathan Barnes, "Aristotle's Theory of Demonstration" *Phronesis* 14 (1969): 123-52; G.E.L. Owen, "TITHENAI TA PHAINOMENA" *Aristote et les Problèms de Methodé* (Louvain, 1961), pp.,83-103; Ingemar Duering, "Aristotle's Method in Biology" ibid.

2
I am following the Oxford translation of A.J. Jenkinson in the APr.

3
Duering, op.cit.

4
Barnes, op.cit., F. Solmsen, *Die Entwicklung der aristotelischen Logik und Rhetorik Neue Philogische Untersuchungen* 4 (Berlin, 1929) rpt. in *Articles on Aristotle*, ed. by Sorabji et al., op.cit.

5
Cf. Duering, *Aristoteles* (Heidelberg, 1966), p. 22.

6
Solmsen, pp. 126ff.

7
This alternative has been suggested by M. Grene, *A Portrait of Aristotle* (Chicago, 1963), pp. 97-103.

8
Ths alternative has been advocated by G. Grote, *Aristotle* 3rd ed. (London, 1883), p. 210; and by Stachowiak, *Rationalismus im Ursprung* (Vienna/New York, 1971), pp. 273-9.

9
And, of course, even an *epi to polu* science can be demonstrated with modal logic.

10
One cannot equate rigorous with demonstrative since both geometry and arithmetic are equally demonstrative while at APo. 87a 35 arithmetic is said to be more rigorous than geometry. Also in the so-called mathematical *De Caelo* (that is mathematical in a broad sense) it is clear that the cosmology discussed is not demonstrative.

11
The place of the *Topics* here has been documented by E. Weil, "La Place de la Logique dans la Pensee Aristotelicienne" *Revue de Metaphysique et de Morale* 56 (1951): 283-315, rpt. in *Articles on Aristotle* vol. 1, ed. by Sorabji et al. op.cit.

12
This claim refers to a 'real' priority concerning the being of the points in question, and is not dependent upon the actual order of composition of the respective texts.

13
This expression is correctly analyzed by W. Wieland, *Die aristotelische Physik* (Gottingen, 1962) 2nd ed. 1970, p. 217 following C. Arpe, *Das ti en einai bei Aristoteles* (Hamburg, 1938), p. 54. The jist of it is that Aristotle takes an expression which asks for the definition of a man, "*ti ēn anthropoi anthropoi einai*: 'what does it mean for a man to be a man?", and then merely drops out the particular content at hand--in this case the *anthropoi*. What we are left with is the structure of a question about essence without reference to any particular content.

14
Allan Gotthelf, op.cit., also points to the regulating effect of the *eidos* in teleological development, though his final position *is different* in several respects (one being that the form is never of Nature but always of nature).

15
This position is taken by J. Rist, "Aristotelian Teleology" *Transactions and Proceedings of the American Philological Association* 96 (1969): 337-49; E. Zeller, tr. as *Aristotle and the Earlier Peripatetics* vol. 1 (London, 1897), p. 459-61; Max Hocutt, "Aristotle's Four Becauses" *Phronesis* 49 (1974): 398.

16
This is argued for by Jonathan Barnes, *Aristotle's Posterior Analytics* (Oxford, 1975), p.22.

17
This position is upheld by C.G. Hempel, op.cit., and C.J. Ducasse, "Explanation, Mechanism and Teleology" *Journal of Philosophy* (1926): 150-55.

18
Cf. n. 45 in Chapter Three.

19
There are several aspects of my position which could be compared to other writers here. One immediately thinks of the work on cybernetics: A. David, *La Cybernetique et L'Humain* (Paris, 1965); G. Klaus, *Kybernetik in philosophischen Sicht* (Berlin, 1961); N. Wiener, *Cybernetics* (Cambridge, Mass., 1948). There are also some similarities with G. Sommerhoff, *Analytical Biology* (London, 1950) and Ernst Mayr, "Teleological and Telenomic, a New Analysis" in R.S. Cohen and M.W. Wartofsky eds. *Methodological and Historical Essays in the Natural and Social Sciences*, vol. 14 of *Boston Studies in the Philosophy of Science* (Dordrecht, Holland/Boston, 1974), p. 98 in my application of systems analysis to biology. This mode of analysis follows closely the work done by Herbert Simon, particularly *The Sciences of the Artificial* (Cambridge, Mass., 1969). But even though there are these works in the same direction, my own interpretation is slightly different from all of them.

20
This metaphorical application is discussed by Larry Wright, _Teleological Explanations_ (Berkeley, California, 1976), pp. 14-17 and Andrew Woodfield, _Teleology_ (Cambridge, 1976), pp. 32-33.

21
Hempel, p. 299.

22
William Dray has criticized this in _Laws and Explanation in History_ (London, 1957).

23
C.G. Hempel, "The Function of General Laws in History" _Journal of Philosophy_ 39 (1942).

24
Cf. Karl Popper, _Logik der Forschung_ (Vienna, 1935), section 12; Hempel, 1942, section 4; P. Caws, _The Philosophy of Science_ (Princeton, N.J., 1965), section 13.

25
This view has been questioned by N.R. Hanson, "On the Symmetry of Explanation and Prediction" _Philosophical Review_ 68 (1959) and I. Schelffler, "Explanation, Prediction and Abstraction" _British Journal for the Philosophy of Science_ 7 (1957).

26
Note that this definition forestalls criticisms that are often leveled that teleological explanations are not explanations proper, but merely inferences _to_ an explanation, cf. R. Cummins, "Functional Analysis" _Journal of Philosophy_ 72 (1975).

27
This point has been made forcefully by R. Eberle, D. Kaplan and R. Montague, "Hempel and Oppenheim on Explanation" _Philosophy of Science_ 28 (1961); H. Fain, "Some Problems of Causal Explanation" _Mind_ 72 (1963); R. Ackermann, "Deductive Scientific Explanation" _Philosophy of Science_ 32 (1965).

28
Of course it is the very definition of "explanation" which is at issue here. Under Hempel's idea of explanation teleology fails, but this says more about the adequacy of Hempel's model

than it does about the value of teleology.

29
This problem was posed by the classic papers of R. Chisholm, "The Contrary-to-Fact Conditional" Mind 55 (1946) and Nelson Goodman, "The Problem of Counterfactual Conditionals" Journal of Philosophy 44 (1947). Since then it has been discussed in innumerable articles and books.

30
Though Aristotle's logic did not contain the form and structure of modern logic, it is lame to rest on this defense. Rather, I will try to incorporate modern logic into my exposition of TE and ME.

31
This distinction is made by Richard Sorabji, "Function" Philosophical Quarterly 14 (1964).

32
This answers an objection that Ernest Nagel made in "Teleology Revisited" in Teleology Revisited and Other Essays in the Philosophy and History of Science (N.Y., 1979), p. 298, against Larry Wright. Nagel says that Wright's analysis requires us to say F(i) iff we know or have good reason to believe that i has been selected to be a part in the organism because it brings about F. This corresponds to Wright's 'F' model which is stronger than the 'T' model which he most staunchly defends. However, Nagel says that this analysis fails because it requires i to be known apart from the system which is impossible and because this sort of "selection" rarely occurs (in evolution, for example, one merely has random mutation). But, on the contrary, one can know i apart from the system--one merely does not have a complete knowledge of i in all relevant circumstances, viz., i in the system. But because of my restriction that all levels and parts be robust, i must, in principle, be known distinctly or else it is not a part. This does not require one to know before hand that i will work in the system; the statement of belief (in the case of conscious actions) will suffice. Secondly, evolutionary change at the level of the genome is random mutation, but the upper level TE is not. It is this which "selects" the species which are to survive.

33 The unity of science hypothesis will be attacked shortly.

34 The literature on robustness is quite large; see especially: Donald Campbell, "Common Fate, Similarity and other Indices of the Status of Aggregates of Persons as Social Entities" *Behavioral Science* 3 (1958): 16-25, Donald Campbell and Donald Fiske, "Convergent and Discriminate Validation by the Multi-Trait Multi-Method Matrix" *Psychological Bulletin* 56 (1959): 81-105, Donald Campbell, "Evolutionary Epistemology" in P.A. Shilpp ed., *The Philosophy of Kark Popper* (LaSalle, Illinois, 1974), vol. 1, pp. 413-463; J. Kim, "Supervenience and Nomological Incommensurables" *American Philosophical Quarterly* 15 (1978): 149-156; Richard Levins, "The Strategy of Model Building in Population Biology" *American Scientist* 54 (1966): 421-431, *Evolution in Changing Environments* (Princeton, N.J., 1968); William Wimsatt, "Robustness and Functional Localization: Heuristics for Determining the Boundaries of Systems and their Biases" in M. Brewer and B. Collins eds., *Knowing and Validating in the Social Sciences: A Tribute to Donald T. Cambell* (San Francisco, 1980).

35 This distinction is made between "What is x good for" and "What does x do" by William Wimsatt, "Teleology and the Logical Structure of Function Statements" *Studies in History and the Philosophy of Science* 3 (1972): 1-80.

36 Aristotle was perhaps a little suspicious of this tack, cf. P.A. 683a 20.

37 This does not refer to a temporal order, but a logical order.

38 This argument has been made by Ernest Nagel, *The Structure of Science* (N.Y., 1961), p. 405ff. and is further argued for in Nagel, 1979. Compare also to R.B. Braithwaite, *Scientific Explanation* (Cambridge, 1953), chapter 10.

39 Paul Oppenheim and Hilary Putnam, "Unity of Science as a Working Hypothesis" *Minnesota Studies in the Philosophy of Science*, ed. by H. Feigl, vol. 2 (Minneapolis, 1959). This paper is representative of those who make this claim.

40
Paul A. Weiss, "The Living System" in _Beyond Reductionism_ ed. Arthur Koestler and J.R. Smythies (Boston, 1969), pp. 3-56.

41
William Wimsatt (1975) and David Hull, _Philosophy of Biological Science_ (Englewood Cliffs, N.J., 1974), ch. 1 in which Hull attacks Kenneth Schaffner's "Approaches to Reductionism" _Philosophy of Science_ 34 (1967).

42
Michael Ruse, _The Philosophy of Biology_ (London, 1973), ch. 9 makes an attack on Nagel on this very point, though his argument takes a slightly different form.

43
This is Weiss's example, p. 12.

44
It is unclear whether Aristotle recognized formalized axiomatics. I think that taken in the rigorous sense of Euclid or the modern example of Hilbert this is probably not the case. See: H. Scholz, "Die Axiomatik der Alten" _Blaetter fuer dertsche Philosophie_ 4 (1930): 259-78 rpt. in Heinrich Scholz, _Mathesis Universalis--Abhandlungen zur Philosophie als Strenger Wissenschaft_ 2ed. (Basel, 1969), pp. 27-44; H.D.P. Lee, "Geometrical Method and Aristotle's Account of First Principles" _Classical Quarterly_ 29 (1935): 113-24 and Ian Mueller, _Philosophy of Mathematics and Deductive Structures in Euclid's "Elements"_ (Cambridge, Mass., 1981).

45
By "such and such" configuration here, I mean, within a given tolerance as set by the parameters established by an enumeration of particular spines. Some standard deviation for the various characteristics of spine could be statistically developed and thus offer the subject, "human spine" with such and such configuration--i.e., within the given parameters, the eidos is fixed. This fixity is evident when comparing "human spine" to "ape spine." That such a comparison is meaningful illustrates the determinate nature of "human spine," and "ape spine."

46
For a possible exception to this see Metaph. 1039a 9.

47
These disputes are still going on today: cf. the various systems of systematics outlined in Chapter Two.

APPENDIX

EVENTS AND MOTION

The further explication of what events and motion represented for Aristotle is important if one wants to better understand how ME operates in his writings. As discussed in 3.3, ME comprises both material explanation and motive explanation. These are understood to be operative when Aristotle describes something as happening 'from necessity.'

Today we speak of an event in two ways: 1. To describe a single causal interaction between objects, and 2. To describe a state of affairs which occurs over a given interval of time. The first notion of an event is completely foreign to Aristotle. This notion of an event requires modern concepts of statics and dynamics which Aristotle did not have.{1} Discussions on this tack generally revolve around the problem of whether or not there is any temporal duration between cause and effect (and if there is what is the temporal relationship).{2} Aristotle cannot even enter such a debate since his notion of a cause is not the same. Thus we will look in vain to this literature if we wish to shed light upon Aristotle's notion of an event.

The latter alternative is much more promising. It is important here to establish what is meant by an "event" or a "process." In this account "event" refers to a sequence of interactions between physical bodies under a single description in which the notions of cause and effect are operative to describe these interactions.{3} We could describe one pool ball hitting another and driving it into the right corner pocket as a physical event. We could (in a Newtonian model) describe the interaction of forces and resultant vectors. These are all descriptions with temporal duration. They describe how one drives a ball toward a pocket. However, the kinds of examples that Aristotle often uses generally take place over a longer span of time (such as building a house G.C. 337b 34). Aristotle did not (or could not) describe short-term reactions between bodies. If he did not understand how bodies interacted with each other to create an outcome, then surely our notion of an Aristotelian event (even in sense #2) must be severely restricted.

Though Aristotle does talk about both logical relations and relations of long temporal duration as examples involving something causing something else (gignetai allo ek allou), he also has something roughly analogous to a modern notion of mechanical interaction in mind too [i.e., the ME or material/mechanical explanation]. In G.A. 724a 29 he talks about an immediate effect which results from a well-defined "cause." In this case he mentions that slander causes abuse and abuse a fight. Each action directly triggers another. This is then compared to the action of semen. It is said to trigger biological reactions in the same way (cf. Poet. 1452a 18-21 where events in a tragedy happen because of [dia] each other not merely after [meta] each other).{4} These passages give expression to a doctrine Aristotle had stated earlier (Metaph. 1023a 26ff. and Phys. 190a 22ff.). It shows that Aristotle did have a recognizable notion of efficient cause which might be loosely characterized as "mechanistic" (for further discussion of this term see 3.33). From this notion that chains of events are causally related, one can infer that Aristotle did indeed have a notion of an event or discrete process (at least in the biological realm). I am not trying to infer that Aristotle had any notion of our modern efficient cause. Indeed, it is clear that he did not have an understanding of statics and dynamics, but this does not exclude the possibility that Aristotle did envision sequences in which an efficient cause acted in a "chain-like" manner to bring about some ultimate effect.

Also important in any description of natural events in Aristotle are his discussions of motion. This task could completely engage an entire treatise, but a few things can be said. Aristotle begins Book Three of the Physics with a strong justification for studying motion: "Since Nature is the principle of motion and change, [and since] it is about Nature we study, then we must understand the nature of motion, for necessarily, to be ignorant of this is to be ignorant of

Nature." Knowledge of motion, then, is a prerequisite for knowledge of Nature itself. It is interesting to note that there are four types of motion mentioned (201a 5), which are changes in: a) substance, b) quality, c) quantity and d) place. Each of these may properly be termed motion or change (*kinēsis*, *metabolē*). The changes referred to in these four categories may refer to the state of something being different at successive temporal intervals due to: 1) change in the object's natural actualization (according to the rules or quasi-laws governing what it is), 2) excess and defect, and 3) the agent/patient relation. Of these the actuality/potentiality distinction must be understood as the most important here. (The latter two simply describe how the actuality/potentiality relation operates.) In Book Two of *De generatione et corruptione* the physical potentialities of matter are described. These explain just how alteration (*alloiosis*), increase and decrease, coming-to-be and passing away, and locomotion might occur from a physical description alone (this corresponds to the ME concept discussed in Chapter Three). It is in this way that a physical understanding of the actuality/potentiality relation can be brought forward and it is precisely this which makes Aristotle's theory explanatory by allowing an independent descriptive principle, that independent power of the matter and the contraries, to be between the *explanans* and the *explanandum*.

Motion is defined as "the fulfillment of what is potentiality, in so far as the potential thing actualizes in virtue of this" (201a 10), and "the fulfillment of the potential when it is actualized and acts not in virtue of itself but qua moveable" (200a 28). Thus in constructing a definitive description of an Aristotelian event, one must include the actuality/potentiality distinction as a primary component. However, with respect to the mechanism of this process one must sharpen how the event exists temporally. In the next chapter Aristotle begins laying the necessary groundwork for this. "Motion is the [process of] actualization in virtue of this [process]; this is brought about by being in contact so that at the same time it is affected"(202a 6). This description can create problems between the mover and what is moved. But this apparent difficulty is overcome by the manner in which we describe the event. Understanding an event properly calls for an exact description to prevent two things from having one and the same actualization (202b 5ff.). It is also important to remember that the various types of motion (viz., locomotion, changes in substance etc.) must be distinguished generically as well (227b 4). This allows one's description of the event to be less ambiguous. This exactness is requisite if the student of nature is to be able to ferret out the structure of one particular action or one species of action. This calls further for a clear delineation of an event as having a beginning, a middle and an end. Such an event is said to be one numerically or essentially (227b 22). Whether or not an event is one in either way is determined by the distinctions between 'that which,' 'that in which,' and 'that during which.' In other words, we need a subject of the event which has to have a descriptive and a temporal context (228a 20). This context must either be one in the way just described or be one as continuous. In either case there has to be a clear indication that the event is "on-going." That is, the event must not have intervals of rest for these descriptions are not of one event but of many (228b 3). Other refinements such as the completeness of the event and the regularity of the event demonstrate that Aristotle has given a great deal of thought to an event theory in which an effect is described as the output of physical interactions of various components in contact during a specified description under rigorous specification. Indeed, such rigor is necessary if one is to understand how teleology and necessity operate in Nature.

But perhaps some might be willing to admit that Aristotle has a notion of an event in the realm of phusis, but deny that he had such a notion in the area beyond the moon. This point is more difficult. In De caelo 269b 18 Aristotle states that the bodies beyond the moon are indeed bodies (somata), but that they do not have the normal physical properties of lightness or of weight. That is, they are of an uncorruptible substance which has a different principle of motion within them (circular as opposed to a complicated nexus of rectillinear motions). Light and heavy are defined in terms of their principles of motion and not in terms of their "mass" as any post-Galilean might expect. What is clear is that the planets are bodies which have a different principle of motion than bodies in "nature" (i.e., those on earth). From this it is evident that there is at least a very primitive notion of an event operating, viz., that of a body moving in circular motion. One can mark a point on a circle and note when the body returns to its starting point. One revolution might be termed a recognizable unit of motion. Certainly this trajectory can be seen as uniform. A body must reach and pass through one point before it reaches another. The source of motion is from the nature of the body itself (D.C. 270a). This is why it is difficult to talk about "mechanical" causes in the super-lunar realm. Because their individual motions are efficiently caused internally it appears that there is no "mechanical" principle operating. This is not the case.

There is some kind of material/efficient cause operating (though it may seem to us to be an implausible one) namely that of movement in virtue of one's material composition. These bodies are composed of aither which gives them their potential for movement.

Aristotle uses a notion of an event when he discusses why the universe must move in a certain direction (Book Five). Now it is true that the principle upon which he bases his argument is a theoretical one, but still he does see the single revolution of the universe as a whole (in this case) as a discrete process or event upon which he is to construct his argument.{5}

The motions of heavenly bodies act in accordance with strict laws in a way analogous to the causative laws on earth (292b).{6} This seems to suggest the same kind of interactions noted earlier which occur in *phusis*. But the super-lunar realm is not in *phusis* and so is different. The motions here are *causa sui* so that one must rather speak about events caused by interaction or relative positions of the bodies themselves as seen from the earth. The events thus become such processes as: retrogression, solar heat, eclipses and the phases of the moon. Now it is true that these events were not very well known at the time of Aristotle. Much ofthe work of astronomers was simply mathematical calculation. One might be led to believe that these occurances were only known in terms of terribly rough tables which merely recorded their dates, but which could make no serious attempt at explanation or prediction. Such a view would certainly seem to tell against the event/process model. However, even though a eclipse is very complicated and was beyond the grasp of Aristotle's power of prediction, he still seems determined to offer "mechanical" explanations why there are more lunar eclipses than solar ones by positing invisible bodies between the earth and the sun (293b 22). Aristotle also describes the process of a lunar eclipse at 297b 27 and the phases of the moon at b 25 as being caused by the spherical shape of the earth. Retrogression (296b) and the efficient cause of solar heat (289a 30 cf. Mete. 341a 19) are also described in terms of efficient causes. This suggests that Aristotle believed that there were a series of events or processes which required explanation (no matter how theoretical). That he saw them as events seems clear by the very fact that he singles them out for discussion. That these events were regular (given adequate data and a correct model) seems to be correct from his statement at 292b on superlunar laws. Thus it seems that both in the sublunar realm (in *phusis*) and in the superlunar realm Aristotle believed in at least some form of an event or process (though the evidence for the latter is not as strong). This means that in the sections on necessity one can view absolute necessity as operating in superlunar events or processes. Its objects come about without exception while in the sublunar realm, conditional necessity operates in natural events "for the most part" (see section 3.31). In both types of events it is important to realize that necessity is due to certain conditions. These conditions are not merely mathematical, but actually refer to material/efficient causal connections inhering in and between the bodies therein.

Notes to the Appendix

1
This has been shown by H. Cateron, La Notion de Force dans le Systѐme d' Aristote (Paris, 1923), pp. 11-22.

2
The best statment of the view for simultaneous causation can be found in Myles Brand, "Simultaneous Causation" in Time and Change ed. Peter van Inwager (Dordrecht, 1980), pp. 137-153. For the opposing view see: T.L. Beauchamp and A. Rosenberg, Hume and the Problem of Causation (Oxford, 1981) and A. David Kline, "The 'Established Maxim' and Causal Chains" PSA Proceedings (1982), forthcoming.

3
In taking an event as a single description, as opposed to all possible states of affairs, I am following Donald Davidson's analysis in "Causal Relations" Journal of Philosophy 64 (1967): 691-703 and "Mental Events" in Experience and Theory ed. by L. Foster and J.W. Swanson (Massachusetts, 1970). I should also note that though I am following Davidson on description I do not necessarily hold his vies on causation. For Aristotle, the distinctions made in contemporary literature such as is made by William Wallace, Causality and Scientific Explanation, vol 1 & 2 (Ann Arbor, Michigan, 1972) and J.L. Mackie, The Cement of the

Universe (Oxford, 1974)--to name just two of the better recent books on the subject--are not cogent because of Aristotle's entirely different understanding of efficient cause. It seems correct to me that Aristotle is not talking about constant conjunction in the Humean sense [alternative (1)] but is rather hazy on how aitia is "responsible," via ME, for some effect or outcome. This issue taken in its historical context has been analyzed by Michael Frede, "The Stoics on Cause" in Doubt and Dogmatism: Studies in Hellenistic Epistemology, Proceedings of a conference held in Oxford, 1978, ed. J. Barnes, M. Burnyeat, M. Schofield (forthcoming Oxford). Whether or not the distinction Frede draws between "is an explanation" and "provides an explanation" (cf. neuter adjective vrs. noun forms), is applicable for Aristotle, it is doubtful that Galen used aitia in this way (K, 9.458.7).

4
For a good argument for this position, see Sorabji (1980), p. 53ff.

5
One should not be too harsh in condemning Aristotle's use of theoretical argument in De caelo. For one thing, acquiring a body of accurate data on astronomy takes a very long time. Also, these objects are further away from us than things on earth so that "there is but little evidence available to our senses to enable us to consider them and all the things we long to know about. We have better means of information, however, concerning the things that perish . . . " (P.A. 641b 25-27).

6
'Analogous' here does not mean 'the same.' In 3.31 the laws of the superlunar realm are shown to be controlled by absolute necessity, while those of phusis are governed by conditional necessity. See also discussions on TE and ME in Chapter Three.

BIBLIOGRAPHY

Biological Texts of Aristotle

A. Historia Animalium

Camus, M. Histoire des Animaux d'Aristote. 2 vols. Paris, 1783. With French Translation.

Aubert, H. and Wimmer, F. Aristoteles Tierkunde. 2 vols. Leipzig, 1868. Text with German translation.

Barthelemy-Saint Hilaire. Histoire des Animaux d'Aristote. 3 vols. Paris, 1883.

Creswell, R. Aristotle's History of Animals. London, 1862. English Translation.

Dittmeyer, L. Aristotelis de Animalibus Historia. Leipzig, 1907.

Louis, Pierre. Aristote, Histoire des Animaux. 3 vols. Paris, 1964-69. Text with French Translation.

Peck, A.L. Aristotle: Historia Animalium. 3 vols. London, 1965-1970. Text with English Translation [volumes 1 & 2 completed].

Thompson, D'Arcy W. Aristotle: Historia Animalium. Oxford, 1910. English Translation.

Tricot, Jules. Aristote, Histoire des Animaux. 2 vols. Paris, 1957. French Translation.

B. *De Partibus Animalium*

Balme, David. *Aritotle De Partibus I and De Generatione Animalium I (With passages from II, 1-3)*. Oxford, 1972. English Translation.

Duering, Ingemar. *Aristotle's de Partibus Animalium. Critical and Literary Commentaries*. Goeteborg, 1943. Commentary.

Langkevel, B. *Aristotelis de Partibus Animalium*. Lipsiae, 1868.

Le Blonde, J.M. *Aristote, Philosophe de la Vie. Le Livre Premier du Traité sur les Parties des Animaux*. Paris, 1945. Text with French Translation.

Louis, Pierre. *Aristote, Les Parties des Animaux*. Paris, 1956. Text with French translation.

Ogle, W. *Aristotle on the Parts of Animals*. London, 1882. revised edition. Oxford, 1912. English translation.

Peck, A.L. *Aristotle: Parts of Animals*. London, 1937. Text with English translation.

C. *De Generatione Animalium*.

Aubert, H. and Wimmer, F. *Aristotelis de Generatione Animalium*. Leipzig, 1860. Text with German translation.

Gohlke, Paul. *Aristotles. Ueber die Zeugung der Geschoepfe (Herausgegeben, ubertragen und erlaeutert)*. Paderborn, 1959.

Louis, Pierre. *Aristote, de la Generation des Animaux*. Paris, 1961. Text with French translation.

Lulofs, H.J. Drossaart. *Aristotelis de Generatione Animalium*. Oxford, 1965.

Peck, A.L. *Aristotle: Generation of Animals*. London, 1943. Text with English translation.

Platt, Arthur. *Aristotle: De Generatione Animalium*. Oxford, 1912. English translation.

D. *Parva Naturalia*.

Beare, J.T. and Ross G.R.T. *Parva Naturalia*. Oxford, 1930. English translation.

Biehl, Guilelmus. *Aristotelis Parva Naturalia*. Lipsiae, 1898.

Hett, W.S. *Aristotle: On the Soul, Parva Naturalia*. London, 1935. Text with English translation.

Lulofs, H.J. Drossaart. *Aristotelis de Somno et Vigilia*. Leiden, 1943. Text with Latin translation.

Aristotelis de Insomniis et de Divinatione per Somnum. Leiden, 1947.

Mugnier, Rene. *Aristote, Petits Traites d'Histoire Naturelle*. Paris, 1965. Text with French translation.

Ogle, William. *Aristotle: On Youth and Old Age, Life and Death, and Rospiration*. London, 1897. English Translation.

Ross, G.R.T. *Aristotle: De Sensu et de Memoria*. Cambridge, 1906. Text with English translation.

Ross, W.D. *Aristotle: Parva Naturalia*. Oxford, 1955.

Siwek, Paulus. *Aristotelis Parva Naturalia*. Rome, 1963. Text with Latin translation.

Tricot, Jules. *Aristote. Parva Naturalia, suivis du traite Pseudo- Aristotelicien de Spiritu.* Paris, 1951. French translation.

E. *De Motu Animalium* and *De Incessu Animalium*.

Farquharson, A.S.L. *De Motu Animalium, de Incessu Animalium.* Oxford, 1912. English Translation.

Forester, E.S. *Aristotle: Movement of Animals, Progression of Animals.* London, 1937. Text with English translation.

Jaeger, W. *Aristotelis de Animalium Motione, de Animalium Incessu, Pseudo-Aristotelis de Spiritus libellus.* Lipsiae, 1913.

Louis, P. *Marche des Animaux et Movement des Animaux.* Paris, 1973.

Nussbaum, Martha. *Aristotle's De Motu Animalium.* Princeton, New Jersey, 1978. Text with English translation.

Torraca, Luigi. *Aristotelis de Motu Animalium.* Naples, 1958. Text with Latin and Italian translations.

E. *De Anima*.

Hamlyn, D.W. *Aristotle's De Anima Books II and III.* Oxford, 1968. English Translation.

Hett, W.S. *Aristotle: On the Soul, Parva Naturalia, On Breath.*

London, 1935. Text with English translation.

Jannone, A. ed. and Barbotin, E. (trans.). Aristotle, de l'Ame. Paris, 1966. Text with French translation.

Ross, W.D. Aristotle: Psychology. Oxford, 1942. English Translation.

Aristotelis de Anima. Oxford, 1956.

Aristotle: de Anima. Oxford, 1961. Text with Commentary.

Siwek, Paulus. Aristotelis Tractatus De Anima. Rome, 1965. Text with Latin translation.

Tricot, Jules. Aristote, de l'Âme. Paris, 1934. French Translation.

General Bibliography

Allan, D.J. The Philosophy of Aristotle. Oxford, 1970 rpt. of 1952.

Anscombe, G.E.M. "Aristotle and the Sea-Battle." Mind 65 (1956): 1-15.

Anton, J.P. Aristotle's Theory of Contrariety. London, 1957.

"The meaning of HO LOGOS TES OUSIAS in Aristotle's Categories 1a." The Monist 52.2 (1968): 252-267.

Ayala, F. "Teleology in Evolutionary Biology." Philosophy of Science 37 (1970): 1-15.

Baker, A.E. "Purpose and Natural Selection: A Defense of Teleology." Scientific Journal of the Royal College of Science 4 (1934): 106-109.

Balme, D.M. "Greek Science and Chance. 1. Aristotle on Nature and Chance." *The Classical Quarterly* 33 (1939): 129-138.

"Greek Science and Mechanism" 2. The Atomists." *The Classical Quarterly* 35 (1941): 123-138.

"Aristotle's Use of Differentiae in Zoology." *Aristote et les Problèmes de Méthode*. Louvain, 1961.

"The Development of Biology in Aristotle and Theophrastus: The Theory of Spontaneous Generation." *Phronesis* 7 (1962): 91-104.

"GENOS and EIDOS in Aristotle's Biology." *The Classical Quarterly* 12 (1962): 81-98.

"Aristotle's Use of Teleological Explanation." Unpublished Inaugural Lecture at Queen Mary's College. London, 1965.

Translation of De Partibus Animalium 1 and De Generatione Animalium 1. Oxford, 1972.

"Aristotle's Biology is not Essentialist." *Archiv fuer Geschichte der Philosophie*. 62.1 (1980): 1-12.

"Aristotle's Natural Teleology." Unpublished Lecture Presented at the Conference on Aristotle's Biology, Marquette University, 1981.

Barnes, Jonathan. "Aristotle's Theory of Demonstration." *Phronesis* 14 (1969): 123-52.

Aristotle's Posterior Analytics. tr. and notes. Oxford, 1975.

The Presocratic Philosophers. London, 1979.

"Proof and Syllogism." in *Aristotle on Science: the 'Posterior Analytics'. ed. E. Berti. Paudua, forthcoming*.

Barnes, J., Schofield, M. and Sorabji, R. eds. *Articles on Aristotle I Science*. London, 1975.

Beare, J.I. *Greek Theories of Elementary Cognition from Alcmaeon to Aristotle*. Oxford, 1906.

Bekker, I. *Aristotles Graece*. Vol. 1. Berlin, 1831.

Berka, K. "Aristoteles und die Axiomatische Methode." Das Altertum 9 (1963): 200-205.

Bertier, J. "Introduction à la Lecture de l'Histoire des Animaux." Recherches sur la Tradition Platonicienne. Paris, 1977. p. 75.

Block, Irving. "Truth and Error in Aristotle's Theory of Sense Perception." Philosophical Quarterly 11 (1961): 1-9.

"Three German Commentators on the Individual Senses and the Common Sense in Aristotle's Psychology." Phronesis 9 (1964): 58-63.

Boas, G. "Aristotle's Presuppositions about Change." American Journal of Philology 68 (1947): 404-413.

Rationalsim in Greek Philosophy. Baltimore, 1961.

Bodenheimer, F.S. "Aristote Biologiste." Universite de Paris, Conference du palis de la Decouverte. Series D, no. 15 (1953), pp. 15-34.

The History of Biology; An Introduction. London, 1958.

Bollack, J. Empédocle. 3 vols. Paris, 1965-9.

Bonitz, Herman. Index Aristotelicus. 2ed. Graz, 1955; rpt. from 1870 ed.

Bourgey, Louis. Observation et Experiénce chez Aristote. Paris, 1955.

Observation et Experiénce chez les Medecins de Collection Hippocratic. Louvain, 1953.

Bourgey, L. and Jouanna, J. eds. La Collection Hippocratique et son role dans l'Histoire de la Médicine. Leiden, 1975.

Boylan, Michael. "Mechanism and Teleology in Aristotle's Biology." Apeiron 15.2 (1981): 96-102.

"Teleology and Necessity in Aristotle." Unpublished Lecutre delivered at the Conference on Aristotle's Biology, Marquette University, 1981.

"The Digestive and 'Circulatory' Systems in Aristotle's

Biology." Journal of the History of Biology 15.1 (1982): 89-118.

"The Galenic and Hippocratic Challenges to Aristotle's Conception Theory." Journal of the History of Biology. (1983), forthcoming.

Burnet, John. Greek Philosophy. London, 1960.

Burnyeat, Myles. "Aristotle on Understanding Knowledge." Aristotle on Science: the 'Posterior Analytics'. ed. E. Berti. Paudua, forthcoming.

Byl, S. "Note sur la Place du Coeur et la Valorisation de la MESOTES dans la biologie d'Aristote." L'Antiquite Classique 37 (1968): 467-476.

Recherches sur les Grands Traités Biologiques d'Aristote: Sources Ecrites et Prejuges. Brussels, 1975.

Cain, A.J. Function and Taxonomic Importance. London, 1959.

Canfield, J.V. ed. Purpose in Nature. Englewood Cliffs, N.J., 1966.

Capelle, W. "Das Problem der Urzeugung bie Aristoteles und Theophrastus und in der Folgezeit." Rheinisches Museum 48 (1955): 150-180.

"Zur Entomologie des Aristoteles und Theophrastus." Rheinisches Museum 55 (1962): 56-66.

Carnap, R. "The Methodological Character of Theoretical Concepts." The Foundations of Science and the Concepts of Psychology and Psychoanalysis (Minnesota Studies for the Philosophy of Science 1), ed. H. Feigl. Minneapolis, 1956. pp. 38-76.

Castiglioni, A. Histoire de la Médecine. Paris, 1931.

Cherniss, Harold. Aristotle's Criticism of Presocratic Philosophy. Baltimore, 1935.

Aristotle's Criticism of Plato and the Academy. Baltimore, 1944.

Chevalier, Jacques. La Notion du Nécessaire chez Aristote et *chez ses Prédécesseurs*. Lyon, 1914.

Clark, Edwin. "Aristotelian Concepts of Form and Function of the Brain." *Bulletin of the History of Medicine* (1963), pp. 1-14.

Clark, E. and Stannard, J. "Aristotle on the Anatomy of the Brain." *Journal of the History of Medicine* 37 (1963): 1-14.

Code, A. "What is it to be an Individual?" *Journal of Philosophy* 75 (1978): 647-648.

Cohen, M.R. and Drabkin, I.E. *A Source book of Greek Science*. Cambridge, Mass., 1958.

Cole, F.J. *Early Theories of Sexual Generation*. Oxford, 1930.

A History of Comparative Anatomy from Aristotle to the 17th Century. London, 1944.

Coleman, W. "Lyell and the 'Reality' of Species: 1830-1833." *Isis* 53 (1962): 325-338.

Cooper, John. "Hypothetical Necessity." ed. Malcolm Schofield and Martha Nussbaum. *Language and Logos: Studies in Ancient Greek Philosophy*. Cambridge, 1982. forthcoming.

"Hypothetical Necessity." Unpublished lecture delivered at the Conference on Aristotle's Biology, Marquette University, 1981.

Cornford, F.M. "Greek Natural Philosophy and Modern Science." in *Background to Modern Science*. ed. J. Needham and W. Pagel. Cambridge, 1938. pp. 3-22.

Cuvier, G. *Histoire des Sciences Naturelles*. Paris, 1841.

Deichgraber, K. *Die griechische Empirikerschule: Sammlung und Darstellung der Lehre*. (Berlin, 1930).

Diels, Herman and Kranz, Walther. *Die Fragmente der Vorsokratiker*. 5th ed. Berlin, 1934-54.

Diller, H. "OPSIS ADELON TA PHAINOMENA." *Hermes* 67 (1932): 14-42.

"Der griechische Naturbegriff." in *Kleine Schriften zur*

antiken Literatur. Munich, 1971.

Kleine Schriften zur antiken Medizin. Berlin, 1973.

Drabkin, I.E. "Notes on the Laws of Motion in Aristotle." American Journal of Philology 59 (1938): 60-84.

Duering, Ingemar. Aristoteles De Partibus Animalium, Critical and Literary Commentary. Goeteborg, 1943.

"Notes on the History of the Transmission of Aristotle's Writings." Goeteborg Hoegskolas Arisskrift 56 (1950): 37-70.

"Aristotle the Scholar." Commentationes in Honorem Edwin Linkomies. Arctos N. S. 1 (1954): 61-67.

"Aristotle and Plato in the Mid-Fourth Century." Eranos 54 (1956): 109-120.

and G.E.L. Owen ed. Aristotle and Plato in the Mid-Fourth Century. Symposium Aristotelicum 1957. Goeteborg, 1960.

"Aristotle's Method in Biology." Aristote et les Problèms de Méthode. Louvain, 1961.

Aristoteles, Darstellung und Interpretation seines Denkens. Heidelberg, 1966.

Dupre, J. "Natural Kinds and Biological Taxa." Philosophical Review 90 (1981): 66-90.

East, S.P. "De la Méthode en Biologie selon Aristote." Laval Theologique et Philosophique 14 (1958): 213-235.

Edelstein, Ludwig. "Aristotle and the Concept of Evolution." Classical Weekly 37 (1943/44): 148-50.

Ancient Medicine. Baltimore, 1967.

Eldredge, N. and Cracraft, J. Phylogenetic Patterns and the Evolutionary Process. New York, 1980.

Eucken, R. Die Method der Aristotelischen Forschung. Berlin, 1872.

Evans, M.G. The Physical Philosophy of Aristotle. New Mexico,

1964.

Farrington, B. "The Greeks and the Experimental Method." Discovery. 18 (1957): 68-69.

Feyeraben, P.K. "Problems of Empiricism." Beyond the Edge of Certainty. ed. R.G. Colodny. Englewood Cliffs, New Jersey, 1965.

"Problems of Empiricism, Part II." The Nature and Function of Scientific Theories. ed. R.G. Colodny. Pittsburgh, P.A., 1970: 275-353.

Against Method. London, 1975.

Forbes, W.T.M. "The Silkworm of Aristotle." Classical Philology. 25 (1930): 22-26.

Fritz, H. von. "NOUS, NOEIN and their Derivatives in Pre-Socratic Philsophy (excluding Anaxagoras) II." Classical Philology. 41 (1946): 12-34.

Furth, Montgomery. "Morphogenesis: Some Elements of Aristotelian Embryology and some Ontological Ramifications." Unpublished Lecture delivered to the Society for Ancient Greek Philosophy, 1975.

"Notes on the Construction of Aristotelian Biological Individuals." Unpublished lecture delivered to the Conference on Aristotle's Biology, Princeton University, 1976.

"Transtemporal Stability in Aristotelean Substances." Journal of Philosophy 75 (1978): 624-646.

Gandt, F. de. "La MATHESIS d'Aristote: Introduction zux Analytiques Seconds." Revue des Sciences Philosophiques et Theologiques 59 (1975): 564-600.

"La MATHESIS d'Aristote: Introduction aux Analytiques Seconds." Revue des Sciences Philosophiques et Theologiques 60 (1976): 37-84.

Gotthelf, Allan. "Aristotle's Conception of Final Causality." Revue of Metaphysics 30 (1976): 226-254.

Granger, Herbert. "Aristotle and the Genus-Species Relation."

The Southern Journal of Philosophy 18.1 (1980): 37-50.

"The Differentia and Per Se Accident in Aristotle." Archiv fuer Geschichte der Philosophie 63.2 (1981): 118-129.

Grene, Marjorie. Portrait of Aristotle. Chicago, 1964.

"Aristotle and Modern Biology." Journal of the History of Ideas 33 (1972): 395-424.

"Is Genus to Species as Matter to Form. Aristotle and Taxonomy." Synthese 28 (1974): 51-69.

"Individuals and their Kind: Aristotelian Foundations of Biology." ed. S.F. Spicker. Organism, Medicine, and Metaphysics. Dordrecht, 1978. pp. 121-136.

Griffiths, G.C.D. "On the Foundations of Biological Systematics." Acta Biotheoretica 23 (1974): 85-131.

Haldane, J.S. The Philosophy of a Biologist. Oxford, 1935.

"Aristotle's Account of Bees 'Dances'." Journal of Helenic Studies. 75 (1955): 24-25.

Hamelin, Octave. La Théorie de l'Intellect d'Après Aristote et ses Commentateurs. Paris, 1953.

Hamlyn, D.W. "Aristotle's Account of AISTHESES in the De Anima." The Classical Quarterly 9 (1959): 6-16.

"KOINE AISTHESIS." The Monist 52 (1968): 195-209.

Harris, C.R.S. The Heart and the Vascular System in Ancient Greek Medicine from Alcmaeon to Galen. Oxford, 1973.

Heidel, W.A. The Necessary and the Contingent in the Aristotelian System. Chicago, 1896.

Hippocratic Medicine, Its Spirit and Method. New York, 1941.

Hennig, Willi. Phylogenetic Systematics. tr by D. Dwight Davis and Rainer Zangerl. Urbanna, Illinois, 1966. revision of Grundzuege einer Theorie der phylogenetischen Systematik. 1950.

Hintikka, Jaako. "Necessity, Universality and Time in Aristotle." _Ajatus_ 20 (1957): 65-91.

"An Aristotelian Dilemma." _Ajatus_ 22 (1959): 87-92.

"The Once and Future Sea-Fight." _Philosophical Review_ (1964): 461-492.

"Aristotelian Science." _Nous_ 6 (1972): 55-69.

Time and Necessity. Oxford, 1973.

Knowledge and the Known. Dordrecht and Boston, 1974.

"Aristotelian Induction." _Revue Internationale de Philosophie_ 34 (1980): 422-440.

Hocutt, M. "Aristotle's Four Becauses." _Philosophy_ 49 (1974): 385-399.

Holwerda, D. _PHUSIS_. Groningen, 1955.

Horne, R.A. "Aristotelian Chemistry." _Chymia_ 11 (1966): 21-27.

Horowitz, Maryanne Cline. "Aristotle and Women." _Journal of the History of Biology_ 9.2 (1976): 186-213.

Hull, David. "The Conflict between Spontaneous Generation and Aristotle's Metaphysics." _Proceedings of the Seventh International Congress of Philosophy_. Vol. 2. Newark, Delaware, 1967.

"Individuality and Selection." _Annual Review of Ecology and Systematics_ 11 (1980): 311-332.

"The Principles of Biological Classification: The Use and Abuse of Philosophy." ed. P.D. Asquith and R. Giere. _Philosophy of Science Association Proceedings 1978_ East Lansing, Michigan, 1981.

"Why Species Can't Evolve." Unpublished lecture delivered to the Conference on Aristotle's Biology, Marquette University, 1981.

Huxley, T.H. "On Certain Errors Respecting the Structure of the Heart Attributed to Aristotle." _Nature_ 21 (1879): 1-5.

Jacobs, William. "Preus on Aristotle's Eide." Nature and System 3 (1981): 115-118.

Jaeger, Werner. "Das Pneuma im Lykeion." Hermes 48 (1913): 29-74.

Aristotle: Fundamentals of the History of his Development. tr. R. Robinson. Oxford, 1934.

Diokles von Karystos. Berlin, 1963.

"Diolkes of Carystus, a new pupil for Aristotle." The Philosophical Review 49 (1940): 393-407.

The Theology of the Early Greek Philosophers. Oxford, 1947.

Joachim, H.H. "Aristotle's Conception of Chemical Combination." Journal of Philology 29 (1904): 72-86.

Johnson, H.J. "Three Ancient Meanings of Matter: Democritus, Plato and Aristotle." Journal of the History of Ideas 28 (1967) 3ff.

Joly, Robert. "La Caractérologie Antique Jusqu'à Aristote." Revue Belge de Philologie et d'Histoire 40 (1962): 5-28.

"La Biologie d'Aristote." Revue Philosophie (1968): 219-253.

Kahn, Charles. "Sensation and Consciousness in Aristotle's Psychology." Archiv fuer Geschichte der Philosophie 48 (1966): 43-81.

Keller, Otto. Das homerische Tiersystem und seine Bedeutung fuer das Zoologische Systematik des Aristoteles. Wiesbaden, 1917.

Koerner, Otto. Das Homerische Tiersystem und seine Bedeutung fuer die zoologische Systematik des Aristotles. Wiesbaden, 1917.

Kirk, G.S. and Raven, J.A. The Presocratic Philosophers. Cambridge, 1957.

Kuehn, C.G. Claudii Galeni Opera Omnia. 22 Vols. Hildesheim, 1965. rpt of 1823.

Kullmann, Wolfgang. Wissenschaft und Methode: Interpretationen zur aristotelischen Theorie der Naturwissenschaft. Berlin, 1974.

Kung, Joan. "Some Aspects of Form in Aristotle's Biology." Nature and System 2 (1980): 67-90.

"Aristotle on Thises, Suches and the Third Man Argument." Phronesis 26 (1981): 207-247.

"Aristotle's De Motu Animalium and the Separability of the Senses." Journal of the History of Philosophy 20 (1982): 65-76.

Lakatos, I. Mathematics, Science and Epistemology, Philosophical Papers. Vol. 2. ed.J. Worrall and G. Currie. Cambridge, 1978.

Le Blonde, J.M. EULOGOS et l'Argument de Convenance chez Aristote. Paris, 1939.

Logique et methode chez Aristote. Paris, 1939.

Aristote, Philosophe de la Vie. Paris, 1945.

"The Biological Bias of Aristotle." The Modern Schoolman 12 (1958): 82-84.

Lesher, J. "Nous in the Parts of Animals." Phronesis 18 (1973): 44-68.

"Perceiving and Knowing in the Iliad and Odyssey." Phronesis 26.1 (1981): 2-25.

Lee, H.D.P. "Geometrical Method and Aristotle's Account of First Principles." Classical Quarterly. 21 (1935): 113-124.

"Place Names and the Date of Aristotle's Biological Works." The Classical Quarterly 42 (1948): 61-67.

Lennox, James. "Aristotle on Genera, Species and 'The More and the Less'." Journal for the History of Biology 13.2 (1980): 321-346.

"Aristotle on the Function of Reproduction." Unpublished Lecture delivered at the Conference on Aristotle's Biology, Marquette University, 1981.

Lerner, Michel-Pierre. La Notion de Finalité chez Aristote. Paris, 1969.

Lesky, Erna. Die Zeugugns-und Vererbungslehren der Antike und ihr Nachwirken. Akademie der Wissenschaften und der Literatur. Mainz, 1950, 19. Wiesbaden, 1951.

Lewes, G.H. Aristotle: A Chapter from the History of Science. London, 1864.

Littré, E. Oeuvres complètes d'Hippocrate. 10 vols. Paris, 1839-61.

Lloyd, A. "Genus, Species and Ordered Series in Aristotle." Phronesis 7 (1962): 57-90.

"Aristotle's Principle of Individuation." Mind 79 (1970): 519-529.

Lloyd, G.E.R. "The Development of Aristotle's Theory of the Classification of Animals." Phronesis 6 (1961).

"Right and Left in Greek Philosophy." Journal of Hellenic Studies 82 (1962): 67-90.

"Experiment in Early Greek Philosophy and Medicine." Proceedings of the Cambridge Philological Society. 190 (1964): 50-72.

Polarity and Analogy. Cambridge, 1966.

"The Role of Medical and Biological Analogies in Aristotle's Ethics." Phronesis 13 (1968): 68-83.

Aristotle: The Growth and Structure of his Thought. Cambridge, 1969.

Early Greek Science: Thales to Aristotle. New York, 1970.

"The Empirical Basis of the Physiology of the Parva Naturalia." Aristotle on Mind and the Senses. ed. with G.E.L. Owen. Cambridge, 1978. pp. 215-239.

"Saving the Appearances." Classical Quarterly. N.S. 28 (1978): 202-222.

Magic, Reason and Experience. Cambridge, 1979.

Locard, Arnould. Histoire des Mollusques dans l'Antiquité. Lyon, 1884.

Lones, T.E. Aristotle's Researches in Natural Science. London, 1912.

Lonie, I.M. "Erasistratus, the Erasistrateans and Aristotle." Bulletin of the History of Medicine. 38 (1964): 426-443.

"Cos versus Cnidus and the Historians." History of Science. 16 (1978): 42-75, 77-92.

Louis, Pierre. "Sur la Chronologie des Oeuvres d'Aristote." Bulletin de l'Association Guillaume Budé N.S. 5 (1948): 91-95.

"Le Traite d'Aristote sur la Nutrition." Revue de Philogie. 26 (1952): 29-35.

"Les Animaux Fabuleux chez Aristote." Revue des Etudes Grecques. 80 (1967): 242-246.

"La Génération Spontanée chez Aristote." Congrès International d'Histoire des Sciences (1968): 291-305.

La Decouverte de la Vie. Paris, 1975.

Mc Keon, Richard. "Aristotle's Cpnception of the Development and Nature of Scientific Method." Journal of the History of Ideas. 8 (1947): 3-44.

Mackie, J.L. "What Re is De Re Modality?" Journal of Philosophy 71 (1974): 551-561.

Manquat, M. Aristote Naturaliste. Paris, 1932.

Mansion, Auguste. Introduction à la Physique Aristotelicienne. Louvain, 1913.

Mansion, Suzanne. Le Jugement d'Existence chez Aristote. Louvain, 1946.

"Les Apories de la Metaphysique Aristotelicienne." Autour d"Aristote. Paris, 1955, pp. 141-179.

Mayr, E. Evolution and the Diversity of Life. Cambridge, Mass., 1976.

"Is the Species a Class or an Individual?" *Systematic Zoology* 25 (1976): 192.

"Evolution." *Scientific American* 239 (1978): 46-55.

Mendelsohn, Everett. *Heat and Life. The Development of the Theory of Animal Heat.* Cambridge, 1964.

Meyer, A.W. *The Rise of Embryology*. Palo Alto, California, 1939.

Meyer, H. *Natur und Kunst bei Aristoteles*. Paderborn, 1920.

Meyer, J.B. *Aristoteles Thierkunde*. Berlin, 1855.

Michael Ephesios. *In Parva Naturalia Commentaria*. ed. Paulus Wendland. Berlin, 1903.

In Libros de Partibus Animalium, De Animalium Motione, De animalibus Incessu Commentaria. ed. Michael Hayduck. Berlin, 1904.

Miller, Harold. "Techne and Discovery in Ancient Medicine." *Transactions of the American Philsophical Association* 86 (1955): 51-62.

Modrak, D. "KOINE AISTHESIS and the Discrimination of Sensible Differences in De Anima 3.2." *Phronesis* 26 (1981): 35-54.

Moraux, Paul. *Les Listes Anciennes des Ouvrages d'Aristote." Autour D'Aristote.* Louvain, 1951.

ed., *Aristoteles in der neueren Forschung*. Darmstadt, 1968.

Morsink, Johannes. "Was Aristotle's Biology Sexist?" *Journal of the History of Biology* 12.1 (1979): 83-112.

Mourelatos, A.D.P. "Aristotle's Powers and Modern Empiricism." *Ratio* 9 (1967): 97-104.

The Route of Parmenides New Haven, Conn., 1970.

Mueller, Ian. "Euclid's *Elements* and the Axiomatic Method." *British Journal for the Philosophy of Science* 20 (1969): 289-309.

"Greek Mathematics and Greek Logic." *Ancient Logic and its*

Modern Interpretations. ed. J. Corcoran. Dordrecht and Boston, 1974. pp. 35-70.

Philosophy of Mathematics and Deductive Structure in Euclid's "Elements". Cambridge, Mass., 1981.

Mure, G.R.G. *Aristotle.* London, 1932.

Needham, J. *A History of Embryology.* New York, 1934.

Nussbaum, Martha. *Aristotle's De Motu Animalium.* Princeton, New Jersey, 1978.

"Saving Aristotle's Appearances." *Language and Logos: Studies in Ancient Greek Philosophy.* ed. with Schofield, Malcolm. Cambridge, 1982.

Nuyens, Fr. *L'Évolution de la Psychologie d'Aristote.* Louvain, 1948.

Owen, G.E.L. "TITHENAI TA PHAINOMENA." *Aristote et les Problèms de Méthode.* Louvain, 1961.

The Platonism of Aristotle. London, 1965.

"Aristotle on Time." *Motion and Time, Space and Matter.* eds. P.K. Machamer and R.G. Turnbull. Columbus, Ohio, 1976. pp. 3-27.

Pagel, Walter. "The Reaction to Aristotle in 17th Century Biological Thought." *Science, Medicine and History, Essays in Honour of Charles Singer.* ed. E.A. Underwood. London, 1953.

William Harvey's Biological Ideas. New York, 1967.

Peck, A.L. "The Connate Pneuma: an Essential Factor in Aristotle's Solution to the Problems of Reproduction and Sensation." *Science, Medicine and History, Essays in Honour of Charles Singer.* ed. E.A. Underwood. London, 1953.

"Introduction" to his translations of the Loeb editions of *The Parts of Animals, The Generation of Animals,* and *The History of Animals.* London, 1937, 1943, 1970.

"Aristotle on Kinesis." *Essays in Ancient Greek Philosophy.* ed. by J. Anton and G. Kustas. Albany, New York, 1971., pp. 478-490.

Petit, G. "Aristote et La Biologie Marine: Les Cetaces avec Remarques sur la Classification des Animaux dans l'Oeuvre du Stagirite." Acts du 11th Congres inst. d'Histoire. 1965. pp. 41-19.

Peyer, Bernard. "Uber die zoologischen Schriften des Aristotles." Gesnerus 3 (1946): 58-71.

Platnick, Norman and Nelson, G. "A Method of Analysis for Historical Biogeography." Systematic Zoology 27.1 (1978): 1-16.

Systematics and Biogeography: Cladistics and Vicariance. New York, 1981.

Platt, Arthur. "Aristotle on the Heart." Studies in the History of Science. ed. Charles Singer. Oxford, 1921.

Plochman, G.K. "Nature and the Living Thing in Aristotle's Biology." Journal of the History of Ideas. 14 (1953): 167-90.

Pouchet, Charles. La Biologie Aristotelicienne. Paris, 1885.

Preus, A. "Aristotle's Parts of Animals 2. 16 659b 13-19: Is it Authentic?" The Classical Quarterly 18.2 (1968): 170-178.

"On Dreams 2, 459b 24-460a 33, and Aristotle's opsis." Phronesis 13.2 (1968): 175-192.

"Aristotle's Nature Uses . . . ." Apeiron 3.2 (1969): 20-33.

"Aristotle's Natural Necessity." Studi Internazionali di filosophia 1 (1969): 91-100.

"Science and Philosophy in Aristotle's Generation of Animals." Journal of the History of Biology 3 (1970): 1-52.

Science and Philosophy in Aristotle's Biological Works. New York, 1975.

"Galen's Criticisms of Aristotle's Conception Theory." Journal of the History of Biology 10.1 (1977): 65-85.

"Eidos as Norm." _Nature and System_ 1 (1979): 79-103.

"Aristotelian Taxonomic Dilemmas." Unpublished Lecture delivered at the Conference on Aristotle's Biology, Marquette University, 1981.

Randall, J.H. _Aristotle_. New York, 1960.

Reymond, A. _History of the Sciences in Greco-Roman Antiquity_. London, 1927.

Rist, J.M. "Some Aspects of Aristotelian Teleology." _Transactions and Proceedings of the American Philological Association_. 96 (1965): 337-349.

Robin, Leon. "Sur la Conception Aristotélicienne de la Causalité." _Archiv fuer Philosophie_ 23 (1910).

Greek Thought and the Origins of the Scientific Spirit. New York, 1928.

Aristote. Paris, 1944.

Rohr, Michael. "Aristotle and Genus." Unpublished lecture delivered to the Conference on Aristotle's Biology, Marquette University, 1981.

Ross, W.D. _Aristotle_. London, 1923.

Sarton, George. "The Discovery of Mammalian Egg and the Foundation of Modern Embryology." _Isis_ 16 (1931): 42-61.

A History of Science. Ancient Science through the Golden Age of Greece. London, 1953.

Schramm, M. "Aristotelianism: Basis and Obstacle to Scientific Progress in the Middle Ages." (review of A.C. Crombie, _Augustine to_ Galileo), _History of Science_ 2 (1963): 91-113.

Simpson, G.G. _Principles of Animal Taxonomy_. New York, 1961.

Singer, Charles. "The Rise of Modern Biology." _Studies in the History and Method of Science_. ed. Charles Singer. Oxford, 1921

"Biology," and "Medicine." in : _The Legacy of Greece_. ed. R.W. Livingstone. Oxford, 1921. pp. 163ff., 201ff.

Greek Biology and Greek Medicine. Oxford, 1922.

A Short History of Anatomy from the Greeks to Harvey. New York, 1957.

From Magic to Science. New York, 1958.

Skemp, J.B. "HULE and HUPODOXE." Symposium Aristotelicum 1957. Goeteborg, 1960. pp. 201ff.

Sober, E. "Evolution, Population Thinking, Essentialism." *Philosophy of Science* 47 (1980): 350-383.

Sokal, R.R. and Sneath, P.H.A. *Principles of Numerical Taxonomy*. San Francisco, 1963.

Numerical Taxonomy. San Francisco, 1973.

Solmsen, Friedrich. "Antecedents of Aristotle's Psychology and Scale of Beings" *American Journal of Philology*. 76 (1955): 148-164.

"The Vital Heat, the Inborn Pneuma and the Aether." *Journal of Hellenic Studies*. 77 (1957): 119-123.

Aristotle's System of the Physical World: A Comparison with his Predecessors. Ithaca, New York, 1960.

"Greek Philosophy and the Discovery of Nerves." *Museum Helveticum*. 18 (1961): 150-167; 169-197.

"Nature as Craftsman in Greek Thought." *Journal of the History of Ideas*. 24 (1963): 473-496.

Intellectual Experiments of the Greek Enlightment. Princeton, New Jersey, 1975.

Sorabji, Richard. "Aristotle and Oxford Philosophy." *American Philosophical Quarterly*. 6 (1969): 129-135.

"Aristotle on Demarcating the Five Senses." *Philosophical Review* 80 (1971): 55-79.

"Aristotle, Mathematics and Colour." *Classical Quarterly*. N.S. 22 (1972): 293-308.

Necessity, Cause and Blame. Ithaca, New York, 1980.

Suppe, F. "The Search for Philosophic Understanding of Scientific Theories." The Structure of Scientific Theories. ed. F. Suppe. Urbanna, Illinois. pp. 2-241.

Tannery, P. Pour L'Histoire de la Science Hellène. Paris, 1887.

Thompson, D'Arcy Wentworth. A Glossary of Greek Birds. London, 1895.

Aristotle as Biologist. Oxford, 1913.

Growth and Form. Cambridge, 1917.

"Natural Science." in The Legacy of Greece. ed. R.W. Livingstone. Oxford, 1921.

A Glossary of Greek Fishes. London, 1947.

Thompson, Wayne. Aristotle's Deduction and Induction. Amsterdam, 1975.

Toulmin, Stephen. The Philosophy of Science. New York, 1960.

Ulmer, K. Wahrheit, Kunst und Natur bei Aristotles. Tuebingen, 1953.

Underwood, E.A. ed. Science, Medicine and History, Essays in Honour of Charles Singer. 2 vol. London, 1953.

Wasserstein, A. "Thales' Determination o the Diameteres of the Sun and the Moon." The Journal of Hellenic Studies 75 (1955):114-116.

"Greek Scientific Thought." Proceedings of the Cambridge Philosophical Society. N.S. 8 (1962): 51-63.

Weiss, Helene. Zufall in der Philosophie des Aristoteles. London, 1935.

Kausalitaet und Zufall in der Philosophie des Aristoteles. Basel, 1942.

"Aristotle's Teleology and Uexkuell's Theory of Living Nature." The Classical Quarterly 42 (1948): 44-58.

Weiss, Paul. Principles of Development. New York, 1939.

Wiley, E.O. *Phylogenetics, the Theory and Practice of Phylogenetic Systematics.* New York, 1981.

Wilkins, Burleigh. "Aristotle on Scientific Explanation." *Dialogue* 9 (1970): 337-355.

Wieland, Wolfgang. *Die Aristotelische Physik* Goettingen, 1962. rpt. 2ed. 1970.

Zeller, Eduard. *Outlines of the History of Greek Philosophy.* 13 ed. New York, 1957.

Zuercher, Josef. *Aristoteles Werk und Geist.* Paderborn, 1952.

General Index

[Concepts, Persons, Greek terms]

INDEX

INDEX

INDEX

INDEX

INDEX

Index Locorum

[Aristotelian Texts by Book and Chapter or by Bekker page number]

INDEX LOCORUM